SpringerBriefs in Electrical and Computer Engineering

More information about this series at http://www.springer.com/series/10059

Ana Paula Pinto Correia
Pedro Miguel Cândido Barquinha
João Carlos da Palma Goes

A Second-Order $\Sigma\Delta$ ADC Using Sputtered IGZO TFTs

Springer

Ana Paula Pinto Correia
CTS/UNINOVA and Department
  of Electrical Engineering
Universidade NOVA de Lisboa
Lisbon, Portugal

Pedro Miguel Cândido Barquinha
I3N/CENIMAT and Department of Materials
  Science
Universidade NOVA de Lisboa
Lisbon, Portugal

João Carlos da Palma Goes
CTS/UNINOVA and Department
  of Electrical Engineering
Universidade NOVA de Lisboa
Lisbon, Portugal

ISSN 2191-8112   ISSN 2191-8120   (electronic)
SpringerBriefs in Electrical and Computer Engineering
ISBN 978-3-319-27190-3   ISBN 978-3-319-27192-7   (eBook)
DOI 10.1007/978-3-319-27192-7

Library of Congress Control Number: 2015958077

Springer Cham Heidelberg New York Dordrecht London

Printed on acid-free paper

Springer International Publishing AG Switzerland is part of Springer Science+Business Media (www.
springer.com)

Foreword

The materials science of thin films and associated fabrication process technologies continue to stimulate new technologically significant application areas related to human-machine interaction. A good case in point is the active matrix display, which relies on a layer of thin-film transistor (TFT) electronics (referred to as the display backplane) to drive the display. The backplane is crucial from the standpoint of speed, resolution, and stability, including instability compensation. An interesting material that has emerged for the backplane is the metal oxide semiconductor. The material is transparent and low-temperature processible making it amenable for layering on plastic or even paper substrates. Fully transparent displays have been demonstrated by leading companies such as LG or Samsung, which are starting to create new application areas such as smart windows for automobiles and buildings and immersive environments. These applications place new demands on the TFT, which now will have to go beyond its standard role as a simple switch to new circuit functions.

This book is an abridged version of the materials science and characterization of oxide TFTs tailored to circuit applications. Following a short introduction, the operating principles of TFTs addressing materials selection are covered in Chap. 2. Processing techniques for TFTs along with materials characterization are addressed in Chap. 3 followed by the theory, operation, and current state of the art of thin-film analog-to-digital converters (ADCs). Implementation considerations are reported in Chap. 5 with emphasis on the comparator and sigma-delta modulator ($\Sigma\Delta M$). The book concludes with future perspectives of materials and ADC architectures.

While the design concepts and circuits demonstrated here are based on metal oxide TFT technology, the design considerations can be adapted to a broader range of materials families that support p-channel transistors.

The book is well written and will benefit the engineering design community, materials scientists, physicists, and chemists who are looking for applications of new materials. The book can also serve as a useful reference for graduate or short courses in universities or industry. The authors are renown in the area of oxide TFTs.

Cambridge, UK   Arokia Nathan
August 2015

Acknowledgments

This work has been funded by FEDER funds through the COMPETE 2020 Programme and National Funds through FCT, Portuguese Foundation for Science and Technology, under the projects "Multifunctional nanoscale oxide materials" (EXCL/CTM-NAN/0201/2012), DISRUPTIVE (EXCL/EEI-ELC/0261/2012), INCENTIVO (EEI/UI0066/2014), and strategic projects (UID/CTM/50025/2013) and (UID/EEA/00066/2013). The work has also received funding from the European Communities 7th Framework Programme under grant agreement i-FLEXIS project (ICT-2013-10-611070) and H2020 Programme under grant agreement ROLLOUT project (ICT-03-2014-644631).

Contents

Symbols

C_i	Gate capacitance per unit area
E_B	Breakdown field
E_C	Conduction band
E_G	Bandgap
f_{co}	Cut-off frequency
F_{in}	Input frequency
F_S	Sampling frequency
g_{DS}	Conductance
g_m	Transconductance
I_{DS}	Drain-to-source current
I_G	Gate leakage current
J	Current density
k	Extinction coefficient
L	Channel length
n	Refractive index
N	Resolution
S	Subthreshold slope
V_{DD}	Positive power supply
V_{DS}	Drain-source voltage
V_{GS}	Gate-source voltage
V_{on}	Turn-on voltage
V_T	Threshold voltage
W	Channel width
κ	Dieletric constant
μ_{eff}	Effective mobility
μ_{FE}	Field-effect mobility
μ_{sat}	Saturation mobility

Acronyms

ADC	Analog-to-digital converter
AFM	Atomic force microscopy
ALD	Atomic layer deposition
AMLCD	Active matrix liquid crystal display
BW	Bandwidth
CEMOP	Center of Excellence in Microelectronics Optoelectronics and Processes
CENIMAT	Centro de Investigação de Materiais
CT	Continuous time
CTS	Centre of Technology and Systems
CVD	Chemical vapor deposition
CMOS	Complementary metal oxide semiconductor
DAC	Digital-to-analog converter
DR	Dynamic range
DRC	Design rule check
DSP	Digital signal processor
DT	Discrete time
EDA	Electronic design automation
EDS	Energy-dispersive X-ray spectroscopy
ENOB	Effective number of bits
ESD	Electrostatic discharge
FET	Field-effect transistor
FFT	Fast Fourier transform
FPD	Flat panel display
IC	Integrated circuit
IGZO	Indium-gallium-zinc oxide
IGO	Indium-gallium oxide
IMO	Indium-molybdenum oxide
IZO	Indium-zinc oxide
JFET	Junction field-effect transistor
LTPS	Low-temperature polycrystalline silicon

LSB	Least significant bit
LVS	Layout versus schematic
MESFET	Metal semiconductor field-effect transistor
MIM	Metal insulator metal
MIS	Metal insulator semiconductor
MISFET	Metal insulator field-effect transistor
MOSFET	Metal oxide field-effect transistor
MSB	Most significant bit
OLED	Organic light-emitting device
OSR	Oversampling rate
PCELL	Parameterized cell
PDK	Process design kit
PECVD	Plasma-enhanced chemical vapor deposition
PFBL	Positive feedback latch
PVD	Physical vapor deposition
RF	Radio-frequency
RIE	Reactive ion etching
SEM	Scanning electron microscopy
SAR	Successive approximation register
SNDR	Signal-to-noise plus distortion ratio
SNR	Signal-to-noise ratio
TCO	Transparent conducting oxide
TEM	Transmission electron microscopy
TFT	Thin-film transistor
TLM	Transmission line method
TSO	Transparent semiconducting oxide
UNINOVA	Instituto de Desenvolvimento de Novas Tecnologias
XRD	X-ray diffraction
ZTO	Zinc-tin oxide
$\Sigma\Delta$	Sigma-delta
$\Sigma\Delta M$	Sigma-delta modulator

Chapter 1
Introduction

1.1 Motivation

The trends of technological market have been changing, giving special relevance to the interaction between consumer and product. Displays are probably the best demonstration of this trend, being deeply integrated in our daily life in televisions (TVs), smartphones, computers, etc. In 2014, iHS Inc. identified seven key technologies for the upcoming generations of TVs, where it is possible to highlight exciting concepts such as curved TVs, quantum dots, and transparent displays [1]. In fact, iHS forecasts a growth of transparent displays to US $87 billion in 2025, representing more than half of the global display market by then [2, 3]. In line with this research trend, some prototypes of fully transparent displays have already been shown by leading display companies such as Samsung, AUO, and LG, triggering the concepts of cars and buildings with smart windows or even wearable electronics using flexible substrates. More recently, Samsung announced the first mirror and transparent organic light-emitting diode (OLED) display panels mainly targeted at personalized shopping and informational browsing [4].

This display evolution is largely dependent on the availability of transparent semiconducting and conducting oxides (TSOs and TCOs) that allow the mass production of circuits or backplanes with millions of transistors [5]. Thin-film transistors (TFTs) are the particular class of field-effect transistors (FETs) for thin-film technologies and as other transistors, TFTs are fundamental building blocks to switch or amplify electronic signals. Materials used in TFTs naturally determine key device parameters, such as mobility, nominal operating voltage ranges, and leakage current. It is noteworthy the relevance of semiconductors and dielectrics, that have been studied for years in these devices. Recently, TSOs such as indium-gallium-zinc oxide (IGZO) have achieved an extreme importance due to the combination of low processing temperature, large area uniformity and excellent electrical performance, typically outperforming a-Si and organics, and comparable to low temperature

© The Author(s) 2016   1
A.P.P. Correia et al., *A Second-Order ΣΔ ADC Using Sputtered IGZO TFTs*,
SpringerBriefs in Electrical and Computer Engineering,
DOI 10.1007/978-3-319-27192-7_1

polycrystalline silicon (LTPS) TFTs. These are attributes that make IGZO TFTs interesting even outside the concept of transparent electronics, as demonstrated by LG on their 5″ curved OLED TVs [6]. Dielectrics are also crucial defining TFT performance and stability. High dielectric constant (high-κ) insulators have been integrated in TFTs improving device performance and reliability, particularly when low processing temperatures are envisaged. The extra capacitance provided by these high-κ materials compensates for the higher density of interface defects, typically associated with low temperature processing.

Circuit integration is naturally the major aim when optimizing a specific transistor technology, with building blocks ranging from inverters to more complex systems as analog-to-digital converters (ADCs). In fact, ADCs as well as digital-to-analog converters (DACs) play a key role in a high variety of fields such as electronic, medical, telecommunications, or other systems that need signal processing, because despite all real signals are analog, it is always desirable to process them, as much as possible, in the digital domain. Successful design, simulation, and fabrication of an ADC is thus a very significant demonstration of the capability of a TFT technology and, consequently, shows the ability to produce fully transparent electronic products.

1.2 Book Organization

Besides this introductory chapter, the book has more five chapters organized as follows:

Chapter 2—thin-film transistors: This chapter reviews the operation principles of TFTs and gives a historical perspective, from patents to oxide TFTs. Some background related to oxide semiconductors and high-κ dielectrics used in these TFTs is also provided;

Chapter 3—oxide TFTs @ FCT-UNL: The fabrication process used in TFTs is described in this chapter, with relevance being given to sputtering and patterning techniques. Thin films and TFTs characterization details are also provided. Results obtained on IGZO TFTs using single and multilayer dielectric structures based on Ta_2O_5 and SiO_2 are discussed. Finally, the IGZO TFT modeling is analyzed;

Chapter 4—analog-to-digital converters: This chapter reviews the basic mode of ADCs operation as well as the most suited architectures for being implemented using TFT technology, giving relevance to the adopted $\Sigma\Delta$ architecture. A review of the state-of-the-art is presented focusing ADCs fabricated using thin-film technologies;

Chapter 5—a second-order $\Sigma\Delta$ ADC with oxide TFTs @ FCT-UNL: In this chapter a strong emphasis is given to the circuit design especially to the comparator and the $\Sigma\Delta$ modulator. Furthermore, simulation results are presented and discussed. Layout considerations in order to avoid undesirable effects during processing/characterization are also discussed;

Chapter 6—conclusions and future perspectives: In the last chapter the main conclusions are highlighted and discussed. Future perspectives related with materials, substrates, etc., in TFTs and new approaches in ADCs' architectures are analyzed.

1.3 Main Contributions

This book offers an overview from TFTs production and characterization to their integration in circuits with a good degree of complexity, combining materials science and electronics.

A second-order $\Sigma\Delta$ ADC using oxide TFTs is implemented in this work (as a standard procedure, it has been assumed that the decimation filter will be implemented in the digital-signal processor (DSP) usually available at system level). The transistors employ a sputtered IGZO semiconductor and an optimized dielectric layer, based on mixtures of sputtered Ta_2O_5 and SiO_2. Transistor performance and stability are tuned by varying the number of layers composing the dielectric.

An a-Si:H TFT model developed by Semiconductor Devices Research Group at Rensselaer Polytechnic Institute is adapted to simulate the behavior of these devices with good fitting to experimental data.

Concerning circuits, a great relevance is given to the comparator, whose operation strongly depends on the TFT performance. Regarding ADCs, the $\Sigma\Delta$ architecture is selected to deal with device mismatch, taking into account the fabrication process variability. ADC simulations show excellent results which are above the current state-of-the-art for competing thin-film technologies, such as organic semiconductors or even LTPS.

A paper titled "Design of a Robust General-Purpose Low-Offset Comparator Based on IGZO Thin-Film Transistors" was originated from this work and presented at the IEEE International Symposium on Circuits and Systems (ISCAS 2015).

References

1. J. Hong, Displays growth opportunities in TVs report 2014 abstract seven key emerging TV technologies. iHS, Technical report, 2014
2. DisplayBank, Displaybank special report transparent display technology & market prospect. iHS, Technical report, 2012
3. S.U. News, Samsung expanding transparent display market with a new 46-inch LCD panel [Online]. Available http://www.samsung.com/us/news/20095 (visited on 18 June 2015)
4. Samsung display introduces first mirror and transparent OLED display panels [Online]. Available http://www.displaydaily.com/press-release/25045-samsung-display-introduces-first-mirror-and-transparent-oled-display-panels (visited on 19 June 2015)
5. S. Lee, S. Jeon, R. Chaji, A. Nathan, Transparent semiconducting oxide technology for touch free interactive flexible displays. Proc. IEEE **103**(4), 644–664 (2015)
6. Lg display introduces next generation display technology at sid 2013 [Online]. Available http://www.lgdisplay.com/eng/prcenter/newsView?articleMgtNo=4916 (visited on 18 July 2015)

Chapter 2
Thin-Film Transistors

Abstract Thin-film transistors (TFTs) are key elements for thin film electronics, being their most significant application the pixel switching elements on flat panel displays (FPDs). Semiconductor materials enabling faster TFTs, such as low-temperature polycrystalline silicon (LTPS) or transparent semiconducting oxides (TSOs), hold the promise of expanding TFT application to gate and data drivers or even full systems-on-panel, for increased reliability and lower production costs.

This chapter is an introductory background and a concise historical perspective related to TFTs. Additionally, taking into account that the devices explored in this work use an oxide semiconductor (indium-gallium-zinc oxide, IGZO) and an high-κ dielectric (based on Ta_2O_5 and SiO_2), a brief overview and historical context regarding TSOs and high-κ dielectrics is also provided.

2.1 TFTs Structure and Operation

A TFT is a field-effect transistor (FET) comprising three terminals (gate, source, and drain) and including semiconductive, dielectric, and conductive layers. The semiconductor is placed between source/drain electrodes and the dielectric is located between the gate electrode and the semiconductor. The main idea in this device is to control the current between drain and source (I_{DS}) by varying the potential between gate and source (V_{GS}), inducing free charge accumulation at the dielectric/semiconductor interface [1].

TFTs can be seen as a class of FETs where main emphasis is on large area and low temperature processing, while metal oxide semiconductor field-effect transistors (MOSFETs) are essentially focused in high performance, at the cost of considerably larger processing temperature. As shown in Fig. 2.1, while in MOSFETs a silicon wafer is used, acting as substrate and semiconductor, TFTs use an insulator substrate, such as glass, that is not an active element for device operation. Furthermore, the operation mode is also different between MOSFETs and TFTs. While the former is based on inversion, the latter relies on accumulation.

Depending on the positioning of layers, four TFT structures are typically considered. They can either be staggered or coplanar (whether drain/source and gate are on opposite or on the same side regarding the semiconductor) and, inside

© The Author(s) 2016
A.P.P. Correia et al., *A Second-Order ΣΔ ADC Using Sputtered IGZO TFTs*,
SpringerBriefs in Electrical and Computer Engineering,
DOI 10.1007/978-3-319-27192-7_2

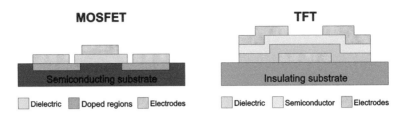

Fig. 2.1 Comparison of typical structures of MOSFETs and TFTs

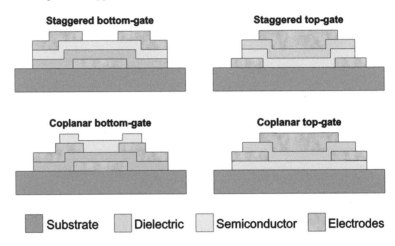

Fig. 2.2 The most typical TFT structures depending on the positioning of layers

them, top or bottom gate (according to the location of gate) [2]. These structures are exhibited in Fig. 2.2. Very briefly, each one has advantages and drawbacks and in terms of fabrication the choice for one of these structures depends on the deposition processes and/or post-processing temperatures or number of lithographic masks involved. For instance, staggered bottom-gate structures are typically used when the dielectric layer requires high temperature, while coplanar top-gate ones are common for high temperature semiconductors, such as poly-Si.

Regarding operation and considering n-type TFTs, these can be designated by enhancement or depletion mode depending if threshold voltage (V_T) is positive or negative. Enhancement mode is typically preferable because a gate voltage is not necessary to turn off the device (to achieve its *Off-state*) [3]. Still, depletion mode devices are also useful for circuit fabrication (e.g., as loads for nMOS logic circuitry).

When $V_{GS} > V_T$, a significant density of electrons is accumulated in dielectric/semiconductor interface and a large I_{DS} starts flowing, depending on the drain-to-source potential (V_{DS}). This state is designated by *On-state* and involves two main regimes depending on the V_{DS} value:

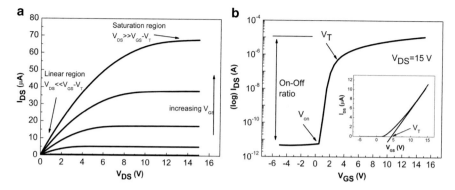

Fig. 2.3 Typical output (**a**) and transfer (**b**) curves for an n-type TFT

- if $V_{DS} < V_{GS} - V_T$, the TFT is in linear/triode mode and I_{DS} is described by:

$$I_{DS} = C_i \mu_{FE} \frac{W}{L} \left[(V_{GS} - V_T)V_{DS} - \frac{1}{2}V_{DS}^2 \right], \qquad (2.1)$$

where C_i is the gate capacitance per unit area, μ_{FE} the field-effect mobility, W the channel width, and L the channel length of the device. For $V_{DS} << V_{GS} - V_T$ the quadratic term is typically neglected.

- if $V_{DS} >> V_{GS} - V_T$, the device is in saturation mode. I_{DS} is independent of V_{DS} and is described by:

$$I_{DS} = \frac{1}{2}C_i \mu_{sat} \frac{W}{L}(V_{GS} - V_T)^2, \qquad (2.2)$$

where μ_{sat} is the saturation mobility.

A typical characterization of TFTs involves static-current voltage measurements where output and transfer curves are obtained as shown in Fig. 2.3. While the output curve provides mostly a qualitative information regarding the effectiveness of channel pinch-off (hence saturation) and contact resistance, the transfer curve offers a more quantitative analysis where some electrical parameters can be determined:

- On–off ratio—This parameter is the ratio of the maximum to minimum I_{DS}. It is known that a higher "on" current offers better driving capability, while a lower "off" current results in low leakage current [2]. Consequently, a higher ratio is preferable;
- Threshold voltage (V_T)—Corresponds to the V_{GS} for which a significant charge is accumulated close to the dielectric/semiconductor interface. A possible methodology to determine this parameter is using a linear extrapolation of the I_{DS}-V_{GS} at low V_{DS};
- Turn-on voltage (V_{on})—Corresponds to the V_{GS} at which I_{DS} starts to increase. It is easily visible in the log I_{DS}-V_{GS} graph, as identified in Fig. 2.3b;

- Subthreshold swing (S)—this parameter indicates the V_{GS} required to increase I_{DS} by one decade, as seen in the subthreshold region. It is defined in V/decade:

$$S = \left(\frac{d\log(I_{DS})}{dV_{GS}} |max \right)^{-1} \tag{2.3}$$

A smaller S is preferable, resulting in lower power consumption and higher speed [2].

Regarding mobility (μ), it directly affects the maximum I_{DS} and the switching speed. There are different methodologies to determine μ, most relevant are highlighted below:

- Effective mobility (μ_{eff})—It is considered as the most correct estimation of μ, which includes the V_{GS} effect. It is determined by the conductance (g_{DS}) at low V_{DS} and requires the previous determination of V_T:

$$\mu_{eff} = \frac{g_{DS}}{C_i \frac{W}{L}(V_{GS} - V_T)} \tag{2.4}$$

- Field-effect mobility (μ_{FE})—It is one of the most used methods to determine μ in TFTs. It is obtained by the transconductance (g_m) at low V_{DS}:

$$\mu_{FE} = \frac{g_m}{C_i \frac{W}{L} V_{DS}} \tag{2.5}$$

- Saturation mobility (μ_{sat})—The determination of this parameter is also common in TFTs and it describes a situation when the effective length is smaller than L [4]. As for μ_{FE}, it is obtained by g_m but at high V_{DS}.

$$\mu_{sat} = \frac{\left(\frac{d\sqrt{I_{DS}}}{dV_{GS}} \right)^2}{\frac{1}{2} C_i \frac{W}{L}} \tag{2.6}$$

The mobility of the free carriers in the channel of device affects directly the maximum operating frequency or cut-off frequency (f_{co}), a parameter extremely relevant to define the possible range of applications of a given TFT technology [4]. It can be defined by:

$$f_{co} = \frac{\mu V_{DS}}{2\pi L^2} \tag{2.7}$$

All these parameters are extremely relevant to evaluate the TFTs performance and understand if they can be integrated into more complex systems.

2.2 An Historical Perspective: From Conceptual Patents to Oxide TFTs

The twentieth century set the birth of electronics, bringing to world concepts as TFTs, integrated circuits (ICs), and complementary metal oxide semiconductor (CMOS) technology, which are of paramount importance nowadays.

The first works on TFTs were reported in 1930 when Lilienfeld described the basic principle, by means of a conceptual patent, of what is known today as metal semiconductor field-effect transistor (MESFET). Some years later, the author introduced the concept of what it nowadays called metal insulator semiconductor field-effect transistor (MISFET) [5]. In the next two decades, two discoveries set the pillars for the modern electronics world: the "Point contact transistor" in 1947 by Bardeen and Brattain and the junction field-effect transistor (JFET) proposed by Shockley in 1952. These were the first transistors that were actually fabricated, showing the switching capability of such devices and how they could be advantageous over the conventional tubes.

In the 60s the first TFT was demonstrated and high speed transistors, the MOSFETs, also emerged in this decade. In 1979 hydrogenated amorphous silicon (a-Si:H) was introduced as a semiconductor on TFTs. Despite its low mobility when compared with the (poly) crystalline materials being studied in that period, the amorphous structure allowed for large area fabrication, which together with the good switching capability of this technology was of great importance in defining a-Si:H TFTs as the main choice for the fabrication of active matrix liquid crystal displays (AMLCDs). Pursuing greater mobility devices, in the 80s, TFTs based on poly-Si were introduced, allowing for high performance circuit fabrication. However, poly-Si TFTs required high temperature processes and had high fabrication cost. Hence, in the 90s, LTPS at around 550 °C was suggested, but processing in large area was still not trivial. Organic TFTs also appeared in this decade with a fantastic advantage, low processing temperature, although their lack of stability and performance still remains an issue these days [5]. Hence, for the new millennium, there was still space for a new technology, combining large area uniformity, low processing temperatures, and good electrical performance. The answer to this emerged in the form of oxide TFTs, which besides these properties also offer the possibility of full transparency.

2.3 Oxide TFTs: Materials, Processes, and Comparison with Other Semiconductor Technologies

Transparent conducting oxides (TCOs) and TSOs, whose studies are reported to the beginning of the twentieth century, are key materials of transparent electronics, exhibiting optical transparency and tunable conductivities between those of conductors and semiconductors [4]. The idea behind them, is to have intrinsic (structural defects) or extrinsic (substitutional elements) doping during and/or after film deposition. For instance, by varying stoichiometry, it is possible to obtain different free carrier concentrations normally in the $10^{21}\,cm^{-3}$ range for TCO and from 10^{14} to $10^{18}\,cm^{-3}$ for TSO [4, 6, 7].

The first work using a TSO (in this case SnO_2) as channel layer in a TFT appeared in the 60s. In the same decade, a work based on ZnO was suggested but with a small I_{DS} modulation by V_{GS} and no I_{DS} saturation was observed [8].

More recently, in the early 2000s, research groups proposed fully transparent ZnO TFTs, produced at 450–600 °C, exhibiting a reasonable performance [3, 9, 10]. After that, it was shown that ZnO could be even sputtered at room temperature without degrading electrical properties of TFTs [11, 12].

In parallel with the "big-boom" of reports on binary compounds as ZnO for TFTs, an initial work by Nomura and co-workers on single crystalline IGZO started an era of incredible success for oxide semiconductors. This multicomponent material produced at 1400 °C exhibited a high μ_{FE} of 80 cm^2/V s when integrated in a TFT structure [13]. However, the more striking aspect was revealed in 2004, when an IGZO layer was fabricated by the same group at room temperature, resulting in an amorphous structure and a TFT with $\mu_{sat} \approx 10\,cm^2/V$ s [14].

What is striking in this a-IGZO and others amorphous TSOs is that they exhibit high μ, not dramatically different from their corresponding single crystals, which is not the case in conventional covalent semiconductors (e.g., a-Si:H typically has $\mu < 1\,cm^2$/V s against $>1000\,cm^2$/V s of Si single crystals). In these ionic materials, the conduction band is made by spherical isotropic ns orbitals of the metallic cations (Fig. 2.4). Hence, if the radii of these orbitals are larger than the distance between cations (verified for $n > 4$), a "continuous path" can be created, improving carrier transport and, consequently, μ [14, 15]. During the last years different combinations of cations have been studied, going from oxides including indium, such as IGZO, indium-zinc oxide (IZO), or indium-molybdenum oxide (IMO) [16–21], generally allowing for high μ_{FE} at lower processing temperatures, to more sustainable approaches (regarding the use of non-critical raw materials) such as zinc-tin oxide (ZTO) [15]. All of them share reasonably high μ (increased with In), low temperature processing, for most of them compatible with low cost flexible substrates, and an amorphous structure that is a great advantage when large area processing is envisaged, as it assures the best possible uniformity.

Fig. 2.4 Schematic for the carries transport path for amorphous oxide semiconductors, proposed by Nomura et al. [14]

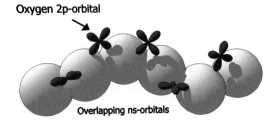

Oxygen 2p-orbital

Overlapping ns-orbitals

2.4 High-κ Dielectrics for Oxide TFTs

Proper choice of a dielectric material is crucial to define the performance/stability of any TFT technology. Alternative dielectrics to SiO_2 have been studied, mainly high-κ as Al_2O_3, HfO_2, and Ta_2O_5. The main advantage of high-κ dielectrics is the possibility to maintain the capacitance per unit area of SiO_2 but with thicker films. This is highly relevant not only for scaling down transistor sizes in c-Si MOSFETs, but also when low temperature technologies, such as oxide TFTs are considered. The importance of this is explained as follows: semiconductor films deposited by lower temperature processes are more prone to have higher densities of defects and reduced compactness, which can be compensated by the larger capacitive injection of high-κ dielectrics. On the other hand, insulators with good capacitance per unit area can still be achieved even if their thickness is increased, compensating the degraded insulating properties of dielectrics fabricated at lower temperatures. The combination of these factors enables low temperature TFTs with low operating voltage, steep subthreshold regions, and large μ, without neglecting the fundamental role of the dielectric as electrical insulator between gate and semiconductor.

High-κ dielectrics typically exhibit a lower bandgap energy, E_G, than the more conventional SiO_2 (Fig. 2.5), which can be problematic in terms of gate leakage current (I_G)—direct tunnelling across dielectric layer, by Schottky emission or Poole-Frenkel effect [22]. Additionally, one has to consider the large E_G of oxide semiconductors when compared to other semiconductor technologies, which turns the high-κ choice for oxide TFTs narrower. Regarding material selection, one still has to consider that the band offset between semiconductor and dielectric should be at least 1 eV. For an n-type device this corresponds to the difference between the minimum of conduction bands of semiconductor and dielectric, while for a p-type device the offset should be analyzed in terms of the maximum of valence bands of both materials [23]. In this way, not all combinations of

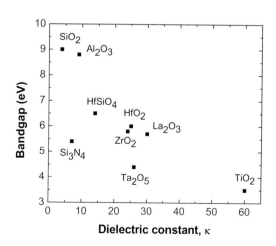

Fig. 2.5 Relation between bandgap and κ for the most relevant inorganic dielectrics

semiconductors and dielectrics are desirable. Regarding material structure, amorphous layers are preferable due to the better uniformity in large areas, smoother dielectric/semiconductor interface and suppression of grain boundaries that act as preferential paths for carriers' flow [24].

Nonetheless and despite the integration of these dielectrics in TFTs is highly promising to improve their performance, a trade-off between κ and breakdown field (E_B) needs to be done [25]. Instability, off-current and hysteresis are seen as a problem, which can be solved using different configurations such as multilayers, or even multicomponent dielectrics, combining high E_G materials as SiO_2, which enables low I_G, high E_B, and good interface properties with most semiconductor technologies, with high-κ ones as Ta_2O_5, HfO_2, or even Al_2O_3, which assure a high capacitance to the overall structure [26–29]. It should also be noted that, as verified with oxide semiconductors, multicomponent dielectrics also potentiate increased structural disorder, hence amorphous structures.

2.5 Current Research Trends in Oxide TFTs

Despite the lack of maturity compared to the dominant TFT technology (i.e., a-Si:H), oxide TFTs are starting to assume a preponderant role in the display industry, with more and more announces of commercially available products integrating this technology, ranging from large area 4k OLED TVs to smartphones. Still, much more progress will certainly be done in the next years on oxide TFTs that will enable an ever more relevant presence in different markets, ranging from fully transparent displays to disposable products. The following paragraphs briefly summarize some of the current research trends in the area.

Research groups are studying new approaches in order to optimize processes, enhance TFT performance and reduce the involved costs. In line with this, there is a constant search for alternative and sustainable oxide semiconductors and dielectrics, both in terms of composing materials (eventually even hybrid inorganic/organic) and structures (where nanostructures as nanowires and nanoparticles are deserving increased attention for ultimate performance levels) [30]. Another important topic is the migration of vacuum processing to simpler solution processing, such as spin-coating or inkjet printing. Although the initial works on solution processed oxide TFTs were based on high temperature processes and yielded low performance [31–33], nowadays there are quite interesting reports on these devices processed at temperatures as low as 200 °C, reaching similar characteristics to their physically processed counterparts [34–36].

In terms of p-type oxide TFTs, required for CMOS architectures using oxide technology, materials as tin and copper oxide have been produced at room temperature and annealed at temperatures as low as 200 °C. However the device performance is far from the one achieved with n-type oxide TFTs [37–39].

As most of these oxide materials can be processed at very low temperatures, and given the great environmental concerns these days, research groups are considering new approaches such as introducing paper in electronics, envisaging a recyclable electronics concept. Flexible substrates and low power circuits have also been considered [24, 40].

Finally, demonstrations of increasingly more complex systems are required to show the potential of this new technology. The ADC presented in this work represents a significant step towards this end, being one of the most complex circuit implementations with oxide TFTs reported so far.

References

1. A.C. Tickle, *Thin-Film Transistors: A New Approach to Microelectronics* (Wiley, New York, 1969)
2. J.-H. Lee, S.-T. Wu, D.N. Liu, *Introduction To Flat Panel Displays* (Wiley, West Sussex, 2008), p. 280
3. R.L. Hoffman, B.J. Norris, J.F. Wager, ZnO-based transparent thin-film transistors. Appl. Phys. Lett. **82**(5), 733 (2003)
4. P. Barquinha, Transparent oxide thin-film transistors: production, characterization and integration. Ph.D thesis, 2010
5. A. Facchetti, T.J. Marks (eds.), *Transparent Electronics* (Wiley, Chichester, 2010)
6. M. Grundmann, H. Frenzel, A. Lajn, M. Lorenz, F. Schein, H. von Wenckstern, Transparent semiconducting oxides: materials and devices. Phys. Status Solidi (a) **207**(6), 1437–1449 (2010)
7. H.Q. Chiang, Development of oxide semiconductors: materials, devices, and integration. Ph.D thesis, Oregon State University, 2007
8. G. Boesen, J. Jacobs, ZnO field-effect transistor. Proc. IEEE **56**(11), 2094–2095 (1968)
9. S. Masuda, K. Kitamura, Y. Okumura, S. Miyatake, H. Tabata, T. Kawai, Transparent thin film transistors using ZnO as an active channel layer and their electrical properties. J. Appl. Phys. **93**(3), 1624–1630 (2003)
10. R.L. Hoffman, Zno-channel thin-film transistors: channel mobility. J. Appl. Phys. **95**(10), 5813–5819 (2004)
11. P.F. Carcia, R.S. McLean, M.H. Reilly, G. Nunes, Transparent ZnO thin-film transistor fabricated by rf magnetron sputtering. Appl. Phys. Lett. **82**(7), 1117 (2003)
12. E.M.C. Fortunato, P.M.C. Barquinha, A.C.M.B.G. Pimentel, A.M.F. Goncalves, A.J.S. Marques, R.F.P. Martins, L.M. Pereira, Wide-bandgap high-mobility ZnO thin-film transistors produced at room temperature. Appl. Phys. Lett. **85**(13), 2541 (2004)
13. K. Nomura, H. Ohta, K. Ueda, T. Kamiya, M. Hirano, H. Hosono, Thin-film transistor fabricated in single-crystalline transparent oxide semiconductor. Science, **300**(5623), 1269–1272 (2003)
14. K. Nomura, H. Ohta, A. Takagi, T. Kamiya, M. Hirano, H. Hosono, Room-temperature fabrication of transparent flexible thin-film transistors using amorphous oxide semiconductors. Nature **432**(7016), 488–492 (2004)
15. T. Kamiya, K. Nomura, H. Hosono, Present status of amorphous InGaZnO thin-film transistors. Sci. Technol. Adv. Mater. **11**(4), 044305 (2010)
16. N.L. Dehuff, E.S. Kettenring, D. Hong, H.Q. Chiang, J.F. Wager, R.L. Hoffman, C.-H. Park, D.A. Keszler, Transparent thin-film transistors with zinc indium oxide channel layer. J. Appl. Phys. **97**(6), 064505 (2005)

17. P. Barquinha, A. Pimentel, A. Marques, L. Pereira, R. Martins, E. Fortunato, Influence of the semiconductor thickness on the electrical properties of transparent TFTs based on indium zinc oxide. J. Non Cryst. Solids **352**(9–20), 1749–1752 (2006)

18. H. Hosono, K. Nomura, Y. Ogo, T. Uruga, T. Kamiya, Factors controlling electron transport properties in transparent amorphous oxide semiconductors. J. Non Cryst. Solids **354**(19–25), 2796–2800 (2008)

19. D. Kang, I. Song, C. Kim, Y. Park, T.D. Kang, H.S. Lee, J.-W. Park, S.H. Baek, S.-H. Choi, H. Lee, Effect of ga in ratio on the optical and electrical properties of GaInZnO thin films grown on SiO/Si substrates. Appl. Phys. Lett. **91**, 910 (2007)

20. P. Barquinha, L. Pereira, G. Goncalves, R. Martins, E. Fortunato, Toward high-performance amorphous GIZO TFTs. J. Electrochem. Soc. **156**(3), H161 (2009)

21. E. Elangovan, K.J. Saji, S. Parthiban, G. Goncalves, P. Barquinha, R. Martins, E. Fortunato, Thin-film transistors based on indium molybdenum oxide semiconductor layers sputtered at room temperature. IEEE Electron Device Lett. **32**(10), 1391–1393 (2011)

22. J Robertson, Interfaces and defects of high-k oxides on silicon. Solid State Electron. **49**(3), 283–293 (2005)

23. J Robertson, B Falabretti, Band offsets of high k gate oxides on high mobility semiconductors. Mater. Sci. Eng. B **135**(3), 267–271 (2006)

24. P. Barquinha, R. Martins, L. Pereira, E. Fortunato, *Transparent Oxide Electronics: From Materials to Devices* (Wiley, Chichester, 2012)

25. J.F. Wager, Transparent electronics. Science **300**(5623), 1245–1246 (2003)

26. L. Zhang, J. Li, X.W. Zhang, X.Y. Jiang, Z.L. Zhang, High-performance ZnO thin film transistors with sputtering $SiO_2/Ta_2O_5/SiO_2$ multilayer gate dielectric. Thin Solid Films **518**(21), 6130–6133 (2010)

27. L. Zhang, H. Zhang, J.W. Ma, X.W. Zhang, X.Y. Jiang, Z.L. Zhang, Copper phthalocyanine thin-film field-effect transistor with $SiO_2/Ta_2O_5/SiO_2$ multilayer insulator. Thin Solid Films **518**(21), 6134–6136 (2010)

28. R.S. Chen, W. Zhou, M. Zhang, M. Wong, H.S. Kwok, Self-aligned top-gate InGaZnO thin film transistors using SiO2/Al2O3 stack gate dielectric. Thin Solid Films **548**, 572–575 (2013)

29. L. Pereira, P. Barquinha, G. Gonçalves, E. Fortunato, R. Martins, Multicomponent dielectrics for oxide TFT, in *Oxide-Based Materials and Devices III*, ed. by F.H. Teherani, D.C. Look, D.J. Rogers. Proceedings of SPIE, vol. 8263 (2012), p. 826316. doi: 10.1117/12.909454. http://spie.org/Publications/Proceedings/Paper/10.1117/12.909454

30. C. Opoku, K.F. Hoettges, M.P. Hughes, V. Stolojan, S.R.P. Silva, M. Shkunov, Solution processable multi-channel ZnO nanowire field-effect transistors with organic gate dielectric. Nanotechnology **24**(40), 405203 (2013)

31. Y. Wang, S.W. Liu, X.W. Sun, J.L. Zhao, G.K.L. Goh, Q.V. Vu, H.Y. Yu, Highly transparent solution processed In-Ga-Zn oxide thin films and thin film transistors. J. Sol-Gel Sci. Technol. **55**(3), 322–327 (2010)

32. M.K. Ryu, K. Park, J.B. Seon, J. Park, I. Kee, Y. Lee, S.Y. Lee, in *AMOLED Driven by Solution-Processed Oxide Semiconductor TFT*, ed. by J. Morreale (Soc Information Display, Campbell, 2009)

33. K. Song, D. Kim, X.S. Li, T. Jun, Y. Jeong, J. Moon, Solution processed invisible all-oxide thin film transistors.J. Mater. Chem. **19**(46), 8881–8886 (2009)

34. Y.-H. Yang, S. Yang, C.-Y. Kao, K.-S. Chou, Chemical and electrical properties of low-temperature solution-processed InGaZn-O thin-film transistors. IEEE Electron Device Lett. **31**(4), 329–331 (2010)

35. K.K. Banger, Y. Yamashita, K. Mori, R.L. Peterson, T. Leedham, J. Rickard, H. Sirringhaus, Low-temperature, high-performance solution-processed metal oxide thin-film transistors formed by a sol–gel on chip process. Nat. Mater. **10**(1), 45–50 (2011)

36. M.-G. Kim, M.G. Kanatzidis, A. Facchetti, T.J. Marks, Low-temperature fabrication of high-performance metal oxide thin-film electronics via combustion processing. Nat. Mater. **10**(5), 382–388 (2011)

37. R. Martins, V. Figueiredo, R. Barros, P. Barquinha, G. Gonçalves, L. Pereira, I. Ferreira, E. Fortunato, P-type oxide-based thin film transistors produced at low temperatures, in *Oxide-based Materials and Devices III*, ed. by F.H. Teherani, D.C. Look, D.J. Rogers. Proceedings of SPIE, vol. 8263 (2012), p. 826315. doi: 10.1117/12.907387. http://proceedings. spiedigitallibrary.org/proceeding.aspx?articleid=1386306
38. B.K. Meyer, A. Polity, D. Reppin, M. Becker, P. Hering, P.J. Klar, T. Sander, C. Reindl, J. Benz, M. Eickhoff, C. Heiliger, M. Heinemann, J. Bläsing, A. Krost, S. Shokovets, C. Müller, C. Ronning, Binary copper oxide semiconductors: from materials towards devices. Phys. Status Solidi B **249**(8), 1487–1509 (2012)
39. R. Martins, A. Nathan, R. Barros, L. Pereira, P. Barquinha, N. Correia, R. Costa, A. Ahnood, I. Ferreira, E. Fortunato, Complementary metal oxide semiconductor technology with and on paper. Adv. Mater. **23**(39), 4491–4496 (2011)
40. R.F.P. Martins, A. Ahnood, N. Correia, L.M.N.P. Pereira, R. Barros, P.M.C.B. Barquinha, R. Costa, I.M.M. Ferreira, A. Nathan, E.E.M.C. Fortunato, Recyclable, flexible, low-power oxide electronics. Adv. Funct. Mater. **23**(17), 2153–2161 (2013)

Chapter 3
Oxide TFTs @ FCT-UNL

Abstract Oxide thin-film transistors (TFTs) optimization is imperative in order to obtain a successful integration of circuits. In fact, parameters as turn-on voltage (V_{on}) or gate leakage current (I_G) are known to influence circuit characteristics. These parameters are greatly affected by the properties of the dielectric layer and its interface with the semiconductor. Therefore, amorphous high-κ dielectrics acquire an important role, especially in multicomponent single or multilayer structures, where materials with different electrical properties (e.g., high-κ and high bandgap energy, E_G) are combined to acquire dielectrics with the best possible performance and reliability.

In this chapter a brief overview about fabrication of thin films and TFTs is provided. Then, it presents a detailed discussion on the characterization of sputtered amorphous multicomponent high-κ dielectrics based on Ta_2O_5 and SiO_2, using single and multilayer structures, and their integration in indium-gallium-zinc oxide (IGZO) TFTs. Finally, an existing model for a-Si:H TFTs is adapted to IGZO TFTs technology.

3.1 Fabrication and Characterization Routes

Several deposition routes are used to fabricate the thin films composing thin-film transistors (TFTs). They are mainly divided into chemical or physical vapor deposition techniques. In the first group, techniques as plasma enhanced chemical vapor deposition (PECVD) and atomic layer deposition (ALD) can be found while in the second electron-beam evaporation and sputtering are the most well-known. Given that all layers of the TFTs in this work were deposited by sputtering, a succinct overview about it is provided here. Furthermore, patterning techniques using standard optical lithography tools were also fundamental for device fabrication and are also briefly explained in this section.

The TFTs were fabricated according to a staggered bottom gate, top-contact structure (Fig. 3.1) on Corning Eagle glass substrates (2.5×2.5 cm). Choice for this structure is based on its simple processing and the advantageous positioning of the dielectric below the semiconductor (sputtering of dielectrics is typically a highly energetic process that can damage the surface of previously deposited layers).

© The Author(s) 2016

A.P.P. Correia et al., *A Second-Order ΣΔ ADC Using Sputtered IGZO TFTs*, SpringerBriefs in Electrical and Computer Engineering, DOI 10.1007/978-3-319-27192-7_3

Fig. 3.1 TFT structure used
in this work: staggered
bottom gate, top-contact

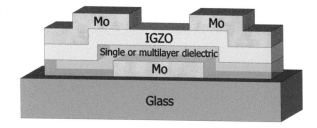

3.1.1   Sputtering

Sputtering is one of the most used techniques to deposit thin films for a wide range of applications, consisting on the bombardment of a source material (target) with ions, being the atoms or molecules removed from the targeted to the substrate where the film grows. When compared with other deposition techniques, sputtering provides some advantages such as good control of thickness and film composition, dense films, good scalability to large areas, and the possibility to deposit a large range of materials, even without intentional substrate heating [1]. As Fig. 3.2 shows, in a vacuum chamber are located two electrodes, the target on top of the cathode and the substrate on the anode (normally grounded, together with the remaining process chamber). When an inert gas such as argon is introduced in the chamber and an electric field is created, electrons collide with gas atoms and depending on the energy transferred during collision electronic excitation or ionization can occur. To come back to the initial state, excited electrons create a "glow discharge" while more electrons and positive gas ions are produced from ionization. These positive ions are directed to the cathode and depending on the energy of ions and the characteristics of the target material, atoms from target are released. The additional electrons promote the "glow discharge" and the continuity of process [1–3]. It is relevant to note that the sputtering yield indicates the number of atoms that are released from target by incident ion, depending both on the incident particle characteristics and on the target material [4].

As previously mentioned, a broad range of materials can be deposited using this technique. Conductive, semiconductor, or even dielectric materials are frequently grown by sputtering with good properties. For this end, different sputtering configurations can be adopted. The simplest one is the so-called DC-diode sputtering where a DC voltage is applied between electrodes. However, this configuration is restricted to conductive targets (insulating materials are not able to maintain the "glow discharge"). On the other hand, on RF sputtering a relatively high frequency AC power source is used (typically 13.56 MHz) and, consequently, the "glow discharge" is preserved, regardless the target electrical conductivity [2, 3]. Nevertheless, in both sputtering configurations mentioned above the deposition rate is low and the electron bombardment can structurally damage the substrate and/or the growing film. To solve these problems magnetron sputtering is used. In this case

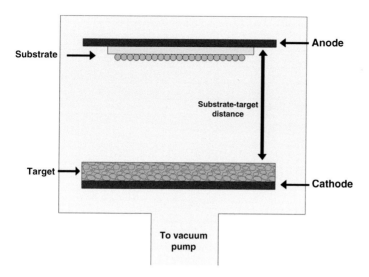

Fig. 3.2 Schematic of sputtering process

a magnetic field is created close to the target, confining electron movement. This reduces substrate bombardment and increases the probability of ionization close to the target, improving the growth rate [1].

Additionally, sputtering can be used with a reactive gas, replacing or adding to the inert gas. This approach, known as reactive sputtering, allows to produce, for instance, oxide thin films such as transparent conducting oxides (TCOs) or transparent semiconducting oxides (TSOs) starting from metallic targets, by using oxygen as reactive gas. Furthermore, reactive sputtering provides a very effective route to control film's stoichiometry, even when using oxide (ceramic) targets [1].

Finally, sputtering systems typically provide the possibility to deposit thin films from multiple targets, either simultaneously (co-sputtering) or sequentially. This allows one to take advantage of the properties of each material in multicomponent and/or multilayer structures.

In This Work RF magnetron sputtering was used to deposit thin films of Molybdenum (Mo, 60 nm thick) as electrodes, IGZO (2:1:2 In:Ga:Zn atomic ratio, 35 nm thick) as active layer and Ta_2O_5/SiO_2-based dielectrics (variable thickness). Deposition conditions for dielectric layer, which is the main focus of this work regarding fabrication process optimization, are shown in Table 3.1. All the depositions were done with a base pressure of $10^{-7}/10^{-8}$ mTorr, without intentional substrate heating and with substrate rotation for improved uniformity.

Different studies were performed regarding the dielectric layer. In some of them, co-sputtering was used with two simultaneous sources SiO_2 and Ta_2O_5, denoted TSiO. The study involved:

Table 3.1 Deposition conditions for different dielectric layers

	Dielectric	
Material	SiO_2	Ta_2O_5
Equipment	AJA ATC-1300F	AJA ATC-1300F
Deposition pressure (mTorr)	2.3	2.3
Target to substrate distance (cm)	18	18
O_2 flow (sccm)	1	1
Ar flow (sccm)	14	14
RF power (W)	150	100
Substrate bias (W)	0–15	0–15
Growth ratio (nm/min)	Depends on material and substrate bias (range: 0.6–2.9)[a]	

All sputtering targets were produced by Super Conductor Materials (SMC, Inc.)
[a] Deposition rates for TSiO (co-sputtered Ta_2O_5 and SiO_2) were 2.5 and 2.9 nm/min, depending if substrate bias was used or not, while for SiO_2 the values obtained were 0.6 and 0.8 nm/min. For Ta_2O_5, the substrate bias did not have significant influence in deposition rate which was 1.9 nm/min

- Thickness variation of single/multicomponent layers (Ta_2O_5, SiO_2 and TSiO), between 150–250 nm and effect of substrate bias[1] during deposition on Ta_2O_5 and TSiO films;
- Multilayer structures based on TSiO and SiO_2 with different number of layers— three (S/T/S) and seven (S/T/S/T/S/T/S);
- Thickness variation of TSiO film when integrated in a multilayer structure (three layers) based on TSiO and SiO_2.

3.1.2 Patterning Techniques

Given that TFTs comprise stacked layers with different materials, patterning techniques are essential to define with a proper resolution the specific pattern of each layer. In this sense, a concise overview about it using standard optical lithography tools is given.

3.1.2.1 Photolithography

Photolithography consists in all the necessary steps to transfer a pattern from a mask to the surface of a substrate, where the photoresist, a material that is sensitive to

---

[1]It was previously shown that, under moderate substrate bias, improved compactness and better insulating properties are achieved, due to re-sputtering of weakly bonded species from the growing film (naturally, this effect is material dependent) [5].

light, plays an important role [4]. Ultraclean conditions are imperative taking into account the resolution (in the case of conventional optical lithography as used in this work around 1 μm), being that any dust particles can compromise the pattern and, consequently, the functionality of the device.

The process starts with substrate cleaning using acetone, isopropyl alcohol, and ultra-pure water. After that, photoresist is applied using spin-coating on the top of a substrate or a film previously deposited. The spin-coating process has been used to produce thin films of different materials (including even oxide semiconductors, as mentioned in Sect. 2.3) and it is a fast and a low cost process. The main idea is to drop a liquid precursor on top of a substrate and spinning the substrate at high speed. The final thickness of photoresist depends on its viscosity and is inversely proportional to the square root of the spinning speed [4].

The substrate with photoresist is then placed on a hot plate to improve the adhesion and to remove by dehydration the solvents of liquid precursor. This step is denominated by soft baking.

After the baking process, the substrate can be aligned with the patterns of a photolithographic mask (typically based on quartz with opaque areas defined by black chromium) in a mask-aligner. When a satisfactory alignment is obtained between alignment marks of the mask and the substrate, the latter can be exposed to the UV light using either contact or proximity mode. Despite bringing more chances of damaging to the mask, contact mode allows to achieve better resolution, as there is no gap for light dispersion between patterns in the mask and photoresist layer. The exposed areas of photoresist to the UV light through the mask have higher solubility than the unexposed areas. Consequently, if the substrate is dipped in a developer solution, the pattern is created by photoresist (Fig. 3.3).

Finally, another baking process (hard baking) is typically required to improve the properties of photoresist. As shown in Fig. 3.3, if the substrate is submitted to an etching process, the thin film directly exposed to the etcher is removed while the areas of film containing photoresist on top are protected. To conclude the process, the photoresist is stripped from the substrate using a resist stripper (e.g., acetone) and a final cleaning process is required. Two additional considerations should be done regarding photolithography:

- There are negative and positive tone photoresists. In the case previously described, a positive photoresist is used, the UV exposed areas are more soluble in the developer. The opposite happens for negative tone resists, i.e., areas exposed to UV get crosslinked and less soluble;
- The masks used in alignment can also be positive or negative (sometimes also designated by bright or dark field). As shown in Fig. 3.3, if positive photoresists and masks (or negative photoresists and masks) are used, etching process is required. Nevertheless, using a negative mask and a positive photoresist, the pattern created with photoresist is the opposite of the final desired pattern. After thin film deposition that covers all area of substrate, if the photoresists is directly stripped by a resist stripper, the thin film on the top of the photoresist areas is also removed, obtaining the desired pattern (Fig. 3.3). This process, known as

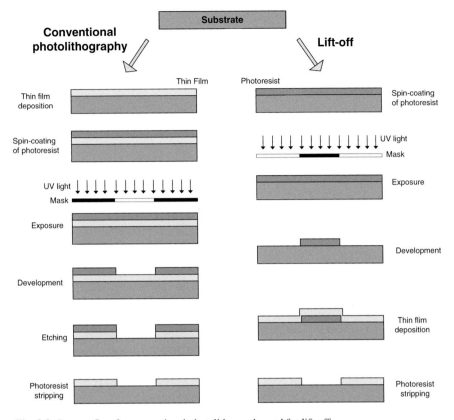

Fig. 3.3 Process flow for conventional photolithography and for lift-off

lift-off, has the advantage of not requiring any etching process (hence there are no selectivity issues), even if resolution is typically worse than the one achieved with etching [1].

3.1.2.2   Etching

The main goal of an etching process is to remove part of a layer unprotected by photoresist from the substrate. This process can be made using a liquid or a gas, denominated by wet or dry etching, respectively. Etching process selection depends on the material that is intended to be removed from substrate and the selectivity required to other layers previously deposited on the substrate [4].

In the case of wet etching, the substrate is immersed in a proper solution (acid or base) that dissolves the film, breaking intermolecular or atomic bonds. This particular etching creates an isotropic profile in films which can compromise the quality of the patterns when high resolution is required, since the material is removed in all the directions [1, 4].

Dry etching is one of the most widely used etching processes in semiconductor manufacturing, allowing for very good resolution and very low amount of residuals. Reactive ion etching (RIE), used in this work, is a particular case of dry etching, being a result of chemical and physical phenomena. Basically, it can be understood as a process similar to sputtering but where the substrate lays in the cathode, i.e., a plasma with an ionized reactive gas bombards the substrate, removing the material. RIE possesses a great flexibility in tuning a process for isotropic/anisotropic etching profile, depending on the gas chemistry and etching parameters [6, 7].

In This Work Process started with substrate cleaning in ultrasonic baths of acetone and isopropyl alcohol and followed by heating it on a hot plate for 20 min at 150 °C, leaving it then to cool down to room temperature. Substrate was then spin-coated with positive photoresist (AZ6612) in a spinner (Headway Research PWM32), firstly at 3000 rpm during 10 s and then at 4000 rpm during 20 s. After that, dehydration process (soft baking) was used—substrate was at 115 °C during 1 min and 15 s on a hot plate. Then, substrate was aligned with a mask[2] using a mask-aligner (Karl Suss MA6) with a 350 W Hg lamp, being the UV exposure performed in soft contact mode during 2.5 s. To obtain the desirable pattern, substrate was introduced in a developer (AZ 726 MIF) for ≈20 s.

After thin film deposition, photoresist and thin film deposited on top of it were removed using acetone (lift-off process), followed by cleaning in isopropyl alcohol and ultra-pure water. A similar method was used for resist stripping after dry etching process of the dielectric layer. Dielectric layer etching was performed in a Alcatel GIR 300 RIE system using as reactive gas sulfur hexafluoride (SF_6). Process parameters were base pressure of 0.05 Pa, gas flow of 10 sccm and RF power of 20 W. The etching rate depended on the material/multilayer structure to be etched, ranging between 20 and 40 nm/min.

3.1.3 Post-deposition Processes

It is well-known that the properties of thin films can be modified by different post-deposition processes, being plasma and annealing treatments some of the most largely used. In the particular case of TCOs and TSOs, post-deposition treatments can largely affect (increase or decrease) the concentration of oxygen vacancies in the structure which has a strong influence in the electrical properties of thin films and devices. Furthermore, these processes can also induce important structural changes in the materials, from small structural rearrangements to amorphous-to-crystalline transitions. On a device oriented perspective, interfaces between different layers can also be significantly affected, e.g., in a transistor the trap state

---

[2]Negative masks were used for lift-off processes of all the layers, except dielectric patterning that required a positive mask for subsequent RIE process.

density in the dielectric/semiconductor interface can be largely affected, influencing important electrical parameters such as V_{on}, subthreshold slope (S), mobility (μ), and stability [1]. Naturally, the main goal with these post-deposition processes is to improve device performance and/or stability, but the exact effect of the process (i.e., increase or decrease electrical conductivity, improve or degrade the quality of a certain interface) heavily depends on the treatment conditions and properties of as-deposited films.

In This Work A broad range of post-deposition processes could be tested for these oxide TFTs, but for the sake of simplicity of the overall process flow a simple post-deposition annealing of the devices at 150–200 °C in air atmosphere during 1 h on a hot plate was performed.

3.1.4 Characterization Techniques

In materials science, a proper characterization allows to study the materials in terms of structural, compositional, morphological, optical, and electrical properties. In the particular case of this work, a coherent characterization of oxide semiconductors and dielectrics thin films is fundamental to define when they have the desired properties to be integrated in devices. Furthermore, the characterization of TFTs containing these oxide semiconductors and dielectrics is relevant to understand the influence of these layers in the general behavior of the device.

Given the main focus of this work on dielectric thin films, these layers were widely analyzed using different characterization techniques which are briefly explained in this sub-section, as well as the nature of information they provide.

3.1.4.1 Structural, Morphological, and Compositional: XRD, AFM, SEM, and RBS

X-ray Diffraction

X-ray diffraction (XRD), a non-destructive technique, is relevant to study structural properties of materials. Mostly, it provides information on whether a material is polycrystalline or amorphous, which is (are) the phase(s) present, the preferential crystallographic orientation, the size, shape, and internal stress of small crystalline regions.

When a X-ray beam reaches a material, the electromagnetic wave intersects the structure and the electrons start to oscillate at the same frequency than the incident beam. As consequence, destructive and constructive interferences of the waves emitted by the atoms are created. The diffracted beam, which is formed by constructive (in phase) interferences, is well-defined for certain angles (θ)

relatively to the scattering planes. The Bragg's law describes the relation between X-ray wavelength (λ), inter-planar space (d), and the diffraction angle [8]:

$$n\lambda = 2d \sin \theta \qquad (3.1)$$

A diffractogram is obtained by varying the diffraction angles (2θ) and analyzing the intensity of diffracted beam. The different intensities depend on the number of atomic planes that the X-ray beam found as equally spaced. These can be compared with a database and the analyzed material and their main properties identified.

In This Work In TFTs, it is extremely relevant to assure that the dielectric layer has an amorphous structure which typically results in a good interface with an amorphous semiconductor and, consequently, in low threshold voltage (V_T) and high field-effect mobility (μ_{FE}). Additionally, suppression of grain boundaries that act as preferential paths for carrier's flow ideally results in lower gate leakage current (I_G) than for a polycrystalline structure of the same material. In this sense, XRD is used to analyze the dielectrics (Ta_2O_5 and SiO_2) in order to define for which range of temperatures the amorphous structure of this layer is preserved.

Atomic Force Microscopy

Atomic force microscopy (AFM) is a technique used to study the materials surface, providing a three dimensional surface profile with atomic level resolution [9]. In comparison with conventional microscopy techniques, AFM provides several advantages, not requiring a vacuum environment or a special sample preparation and being compatible with virtually all kinds of materials.

In terms of operation, one of the most used scanning probe modes is the contact mode that utilizes a sharp tip that scans the sample under the action of a piezoelectric actuator. The tip is fixed to a low spring constant cantilever and it is pushed against the sample due to the low force (similar to the interatomic force range) that is maintained on the cantilever. The repulsive force between the tip and the sample or the tip deflection are registered relatively to spatial variation and then data is converted into the surface image. One possibility to measure the cantilever deflection is using a laser beam that is reflected from the top cantilever surface into a photodiode [9, 10].

Despite contact mode has been widely used, when this mode is utilized in softer surfaces (even liquids), the force on the tip can damage the sample. In this sense, other imaging modes are available including tapping mode. In this mode, while the tip is scanning the sample, the tip-cantilever assembly is oscillating which means that the tip only touches on the sample at the bottom of each oscillation cycle [11].

In This Work AFM (tapping mode) is used in order to study the topography of the dielectric surface and determine its roughness, which is extremely relevant, given that in TFTs the most important operation phenomena take place at the dielectric/semiconductor interface.

Scanning Electron Microscopy

In its essence, scanning electron microscopy (SEM) is used to examine the surface of materials with high resolution and high depth of field. Recent tools can be equipped with a wide range of detectors that greatly expand analytical possibilities, covering morphology, structure, elemental, and even electrical analysis of a multitude of samples, from biological cells to electronic circuits.

It is known that when an energetic electron beam is directed to a sample, interactions such as absorption of electrons by the sample, reflection of primary electrons, emission of secondary electrons, or emission of electromagnetic radiation can occur. In SEM, a low energy beam of electrons (0.5–40 keV) is directed to the sample and the interactions mentioned above are used to form an image by scanning the sample surface and recording the intensities in different detectors. In a conventional SEM, a secondary electrons detector is used and it provides relevant information about topography of the sample. In fact, the number of secondary electrons emitted is dependent on the angle formed between the beam and the peculiarities of surface sample, giving a surface topography analysis. Furthermore, taking into account that backscattered electrons emitted from heavy elements appear brighter in the image than the light elements, a chemical composition analysis can be obtained with a suitable detector [11].

Energy dispersive X-ray spectroscopy (EDS) is used to complement the information provided from SEM, adding elemental analysis to the tool. Basically, EDS uses the incident beam to eject an electron from one of the inner shells, creating a vacancy. Consequently, an electron from an outer shell tends to occupy it. The energy difference between shells is a characteristic of a chemical element and can be released in the form of an X-ray collected by the detector. This method can also provide quantitative information taking into account the peak-area ratios and comparison with material standards.

SEM can also be combined with a focused ion beam (FIB), with the later providing nanoscale milling capabilities. Hence, SEM-FIB provides a valuable tool for failure analysis, allowing to image cross sections of samples at very precise positions, and for transmission electron microscopy (TEM) sample preparation.

Some samples need prior preparation before SEM analysis, specially insulating samples that have to be coated with a thin (few nm thick) conducting layer eliminating the electrical charge-up effect in the sample [4].

In This Work SEM-EDS complements the XRD analysis, not only for a quick evaluation of surface morphology (which can be then studied in more detail with AFM), but most importantly to analyze the atomic ratio of the multicomponent films, as well as eventual contaminations that might arise due to the fabrication process. Furthermore, SEM-FIB is used to do images of cross sections near the device channel.

Rutherford Backscattering Spectroscopy

Rutherford backscattering spectroscopy (RBS) is a technique widely used in materials to determine the elements present in the samples, its stoichiometry and its depth distribution. This spectroscopy technique has two significant advantages over other techniques, offering high precision and not requiring the usage of external standards [12]. Furthermore, computer programs have been develop to analyze and interpret data, which could, otherwise, be very time consuming, particularly for multilayer/multicomponent samples.

Regarding operation, it consists in the bombardment of the sample with high energy ions, typically 0.5–4 MeV. The backscattered ions and the energy distribution is recorded by an energy sensitive detector, at a specific angle. For each element, the backscattering cross section is known which means that it is possible to analyze the sample in a depth profile in terms of composition [13].

In This Work Single and multilayer co-sputtered dielectrics (based on Ta_2O_5 and SiO_2) are analyzed by RBS in order to determine the stoichiometry (Ta_2O_5/ SiO_2 ratio) and its depth distribution.

3.1.4.2 Optical: Spectroscopic Ellipsometry

Spectroscopic Ellipsometry

Spectroscopic ellipsometry is a non-invasive technique which allows measuring changes in the polarization state of light when it is reflected by a surface/sample, analyzing amplitude and phase variation. It can be used to obtain relevant information mainly about composition, film thickness, and surface roughness, which can be all related to optical properties.

It is known that the polarized light can be divided into two components, parallel and perpendicular relatively to the incident plan and, when the light beam is reflected from a surface, it does not have necessarily the same phase. Hence, it is possible to determine the differential changes in phase and amplitude by the Fresnel coefficient (ρ_F):

$$\rho_F = \frac{R_p}{R_s} = \tan \psi \cdot \exp^{j\Delta}, \tag{3.2}$$

where R_p and R_s are the parallel and perpendicular reflection coefficients, respectively, and ψ and Δ are ellipsometric angles. The first angle represents the differential change in amplitude while the second the change in phase [14]. In this spectroscopy method it is necessary to construct an optical model. The components of this model depend on the nature of the material (e.g., Tauc-Lorentz dispersion formulas are typically required for insulators, while TCOs often make use of Drude models to take into account the contribution of free carriers).

In This Work Spectroscopic ellipsometry is used to determine two relevant optical parameters, refractive index (n) and extinction coefficient (k), for single and multilayer dielectric structures. As a consequence, the compactness and the Ta_2O_5 concentration in thin films are evaluated using the Tauc-Lorentz dispersion formula.

3.1.4.3 Electrical Characterization of Dielectric Structures and Thin-Film Transistors

Electrical characterization of devices plays a relevant role in this work to evaluate the success of thin films integration in IGZO TFTs and to define, based on device performance, the constrains needed for circuit design. For this end, static-current voltage characteristics and stress measurements are used in these devices. Additionally, given that a large emphasis is given to dielectric tuning, characterization of metal-insulator-metal (MIM) structures is also imperative.

Evaluation of MIM Structures

Capacitance *versus* voltage (C–V) and capacitance *versus* frequency (C–f) measurements are highly relevant to characterize thin films intended to be used as dielectrics in electronic devices. When metal-insulator-semiconductor (MIS) structures are employed, these techniques are some of the most powerful tools to evaluate important aspects of materials and interfaces such as their defect state densities, but even the analysis with simpler MIM structures provides very useful data, such as capacitance of the dielectric that is then used to access the dielectric constant. By measuring this capacitance over different frequencies, fast or slow polarization mechanisms can be easily traced.

A complete assessment of fundamental properties of a dielectric layer would not be completed without knowing its breakdown field, which can be easily done with an I–V measurement in a MIM structure.

In This Work C–V, C–f, and I–V measurements are performed to access fundamental electrical characterization of MIM structures with different multicomponent/multilayer dielectrics and to infer about the viability of integrating them in TFTs. Parameters extracted were capacitance and breakdown field. Measurements are all done in the dark, at room temperature, and in air.

Static-Current Voltage Characteristics

The measurement of output and transfer characteristics provides all the required information to access static performance of TFTs. Output characteristics mostly provide a qualitative evaluation, giving for instance information about if saturation is achieved (mandatory for circuit application) or if there is a significant contribution of contact resistance (considering the drain-to-source current, I$_{DS}$, at low drain voltage, V$_{DS}$, region). On the other hand, transfer characteristics generally provide a more quantitative analysis, being possible to extract important parameters such as on–off ratio, V$_{on}$, V$_T$, and μ$_{FE}$ [15].

In This Work Transfer and output curves are measured for IGZO devices with different dielectric structures and different annealing temperatures for electrical parameters extraction. Transfer plots are taken both in linear and saturation regimes. It is noteworthy that all measurements are done in the dark, at room temperature, using a semiconductor parameter analyzer and a probe station.

Stress Measurements

Stress measurements provide information about the degradation mechanisms of the transistors and consequently if they are suitable for integration in circuits. Constant gate voltage is one of the most common methodologies, where the basic idea is to apply a constant voltage to the gate electrode and maintain the drain and source electrodes grounded. This way, the conductive channel close to the dielectric/semiconductor interface is forced to be formed, being investigated if over time there is any temporary or permanent degradation on device performance. For this end, stress and recovery (i.e., time elapsed after stress without any stimulus being applied) periods are quickly interrupted, from time to time, to extract the TFTs transfer curves [1].

In This Work A positive gate-bias stress is applied in TFTs and the electrical properties analyzed mainly in terms of V$_{on}$ and S variation during stress and recovery.

3.1.4.4 Summary of Characterization Techniques Used

The most relevant details and information about equipment used during characterization of thin films and TFTs are summarized in Table 3.2.

3.2 Amorphous Multicomponent High-κ Dielectrics Based on Ta$_2$O$_5$ and SiO$_2$: Thin Films and Integration in IGZO TFTs

On the materials side, optimization of sputtered dielectrics based on Ta$_2$O$_5$ and SiO$_2$ was the fundamental part of this research work. Choice for Ta$_2$O$_5$ as the base high-κ material is justified by its amorphous structure (even before mixed with SiO$_2$), high-κ (≈25) and high sputtering rate, which is quite hard to achieve

Table 3.2 Main goals and experimental details of characterization techniques used in this work

Characterization technique (equipment used)	Main goal	Most relevant experimental details
X-ray diffraction PANalytical X'Pert PRO MRD (MRI temperature Chamber)	The onset of crystallization of dielectric films	Temperature range: 100–900 °C Temperature step: 100 °C (three scans at each temperature)
Atomic force microscopy (Asylum MFP-3D)	Topography of the dielectric surface Surface roughness	Area: $2 \times 2 \ \mu m^2$ Mode: Tapping
Scanning electron microscopy (Energy dispersive X-ray spectroscopy) (Zeiss Auriga CrossBeam Workstation)	Atomic ratio of the multicomponent films and eventual contaminations	EHT :15.00 kV Aperture size :60 μm
Rutherford backscattering spectrometry (Van de Graff accelerator (2.5 MV) 2 MeV alpha particle He$^+$ beam)	Ta_2O_5/SiO_2 ratio and relative density of TSiO film and multilayer structures	Si barrier detector: at 165° Incident angle of beam in sample: 10° Current: 2–3 nA Cumulative charge: 3 μC Simulated annealing algorithm: WiNDF [12]
Spectroscopic ellipsometry (Jobin Yvon UVISEL DH-10)	Compactness and band gap of the Ta_2O_5 and TSiO (multi)layers	Modulation software: DELTAPSI Energy range: 1–6 eV
Electrical characterization (Agilent 4155C, Cascade Microtech M150, Keithley 4200SCS and Janis ST-500)	Transfer and output characteristics Stress measurements C–V and C–f plots Parasitic capacitances	Stress with gate field of 0.16 MV/cm Maximum frequency used in C–V and C–f: 1 MHz
Profilometry (Ambios)	Thickness of the dielectric thin films	Force (tip): 1.0 mg Speed: 0.10 mm/s

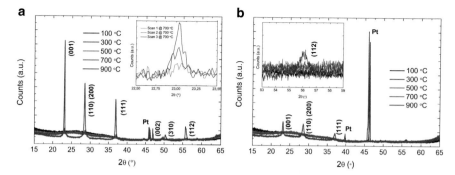

Fig. 3.4 Diffractogram for Ta$_2$O$_5$ (**a**) and TSiO (**b**) thin films, annealed at different temperatures until the crystallization is evident. The data obtained is compared with the ICSD database and corresponds to orthorhombic β-Ta$_2$O$_5$ (ICSD: 98-004-8854)

with sputtered dielectrics. Experiments involved both single and multilayers, with varying thicknesses, effect of substrate bias and influence of the number of layers in multilayer stacks.

Given that the primary focus of the global work was on device optimization and circuit design, the traditional route of material's optimization followed by device integration was not followed here, being both routes advanced in parallel. Still, for clarity of presentation, material properties are presented first, followed by TFT integration.

3.2.1   Single Layer Structure Using Ta$_2$O$_5$ and TSiO

As pointed out earlier, dielectrics with amorphous structure are preferred for generic large area fabrication due to improved uniformity and, in particular, for TFT application, for enhanced performance and reliability. Although this experimental work on TFTs is limited to 200 °C (in order to assure compatibility with a broad range of plastic substrates), XRD analysis of the fabricated dielectrics annealed at higher temperatures allows for a better understanding of the studied materials systems, defining for which range of temperatures they can preserve the amorphous structure.

Two thin films of Ta$_2$O$_5$ and TSiO on Si deposited by sputtering using substrate bias were analyzed between 100 and 900 °C with temperature steps of 100 °C and doing three consecutive scans at each temperature step. In Fig. 3.4, the structural data obtained for these films is shown (curves are presented in steps of 200 °C to simplify the analysis). As expected for these dielectrics, only a very broad peak

a   **b**

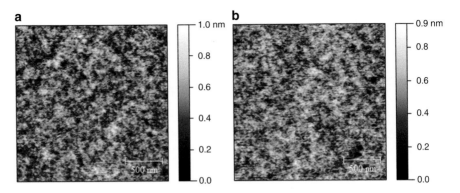

Fig. 3.5 AFM analysis, showing the amorphous structures of Ta_2O_5 (**a**) and TSiO (**b**) thin films

near $2\theta = 25°$ is visible[3] confirming the amorphous structure of both materials. For Ta_2O_5, Fig. 3.4a, some peaks start to appear at 700 °C, being more pronounced at 900 °C and attributed to β-Ta_2O_5. In the inset of this figure, the three consecutives scans for 700 °C are shown which reinforce the idea that the crystallization starts at this temperature by seeing the increase of the intensity of the peaks in each scan. For TSiO, Fig. 3.4b, it would be predictable a higher crystallization temperature due to incorporation of SiO_2—mixing these materials a structural disorder is induced and, in fact, crystallization only occurs at 900 °C. Both crystallization temperatures are higher than the thermal budget considered in this work, meaning that amorphous structure is preserved for all the fabricated TFTs.

The amorphous structure was also suggested by AFM, as shown in Fig. 3.5. It is possible to note that surface of both thin films is completely smooth, with RMS roughness values close to detection limit of equipment, below 0.5 nm.

An elemental analysis of TSiO thin films was done using RBS technique in Nuclear and Technological Institute (ITN), Portugal. The results obtained were fitted, fixing the oxides composition and allowing the system to adjust the stoichiometry $(Ta_2O_5)_x (SiO_2)_y$. According to the results, the thin film presents a Ta_2O_5/SiO_2 ratio of 2.2 and an areal density of 1.4×10^{18} at/cm². The thin film thickness measured by profilometry was 225 nm, which means that considering the thickness obtained by RBS, the co-sputtered material presents a density near 6.2×10^{22} at/cm³. Considering the molar mass (441.9 g/mol), the density of β-Ta_2O_5 (8.2 g/cm³) and the number of atoms present in each molecule of Ta_2O_5 (7 atoms), the density of Ta_2O_5 is 7.8×10^{22} at/cm³. The density obtained for the TSiO film shows a good match to the theoretical density of Ta_2O_5, suggesting a structure with good compactness.

---

[3]In both dielectrics a peak at $2\theta = 46°$ is visible and is due to the Pt foil (the heating element) where the sample is mounted.

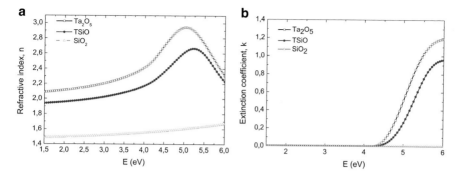

Fig. 3.6 Spectroscopic ellipsometry analysis: comparison between Ta$_2$O$_5$, TSiO, and SiO$_2$ thin films in terms of refractive index (**a**) and extinction coefficient (**b**)

Table 3.3 Properties of Ta$_2$O$_5$ and TSiO films obtained by Tauc-Lorentz dispersion method

Sample	Thickness (nm)	Roughness (nm)	E$_G$ (eV)
Ta$_2$O$_5$	267.4	1.1	4.20
TSiO	237.7	0.4	4.23

The spectroscopic ellipsometry technique acquires a high importance in dielectrics study and, consequently, thin films of Ta$_2$O$_5$ and TSiO on Si were analyzed using it. The Tauc-Lorentz dispersion formula was used and the best fitting results were obtained for two oscillators. In Fig. 3.6, dependences of refractive index (n) and extinction coefficient (k) on energy (E) are shown and, in Table 3.3, some properties obtained by Tauc-Lorentz dispersion formula are presented.

The positioning of the TSiO plots in Fig. 3.6 clearly suggests that SiO$_2$ is incorporated in TSiO, although the Ta$_2$O$_5$ concentration is considerably higher (TSiO plot closer to Ta$_2$O$_5$ in terms of values and shape). These results are in agreement with RBS analysis shown before. It is noteworthy that due to the usage of substrate bias during thin film deposition, the percentage of SiO$_2$ in film is lower, as substrate bias preferentially re-sputters Si atoms from the film [5]. Furthermore, the broad peak in n-E plot suggests that samples have an amorphous structure, in agreement with previous results of XRD and AFM. In terms of relative density, the refractive index for TSiO is lower indicating that its density is lower comparing to Ta$_2$O$_5$, which is consistent with RBS analysis. Regarding values of energy that induce absorption in the beginning of conduction band, these are shown in Fig. 3.6b. E$_G$ is slightly higher for TSiO than for Ta$_2$O$_5$ film, suggesting that SiO$_2$ is present in film but its band structure remains very similar to Ta$_2$O$_5$. On the other hand, the E$_G$ value even for Ta$_2$O$_5$ is lower than the typically reported (4.5 eV), most likely due to the band tails arising from the physical damage of the sputtering process, taken into account by the model [5]. In terms of roughness, the values obtained indicate a flat surface in accordance with AFM images.

Given that the main intent is to integrate these dielectrics in TFT structures, it is imperative to analyze several electrical parameters such as κ, breakdown field (E$_B$), capacitance per unit area (C$_i$), and current density (J). For this end,

Table 3.4 Electrical properties of Ta$_2$O$_5$ and TSiO determined from MIM structures

Sample	Thickness (nm)	E$_B$ (MV/cm)	κ	C$_i$ (nF/cm^2)	J @ E$_{Bmax}$ (A/cm^2)
Ta$_2$O$_5$	230.0	1.54	22.6	87.0	1.21×10^{-2}
TSiO	210.0	2.81	17.0	71.7	8.88×10^{-5}

Results for SiO$_2$ films are not presented in this table, given that for the deposition conditions used herein these films always revealed very large leakage current, inhibiting proper parameter extraction

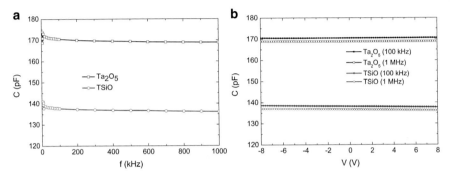

Fig. 3.7 C–f and C–V characteristics of MIM structures that integrated Ta$_2$O$_5$ and TSiO layers (thickness $\approx$220 nm)

MIM structures annealed at 150 °C were analyzed, being the results summarized in Table 3.4 and the C–f and C–V curves presented in Fig. 3.7. It is possible to observe that TSiO results in a lower C, given its lower κ and that there is no significant variation when capacitance is measured using higher frequencies. This is a good indication that within the range of frequencies analyzed slow polarization mechanisms are not relevant. Furthermore, a trend for higher κ for dielectrics with higher Ta concentration is verified. In fact, κ obtained for Ta$_2$O$_5$ is extremely close to the values reported in literature ($\approx$25) [16]. TSiO exhibits a higher E$_B$ than Ta$_2$O$_5$ which is highly relevant to improve TFT and circuit reliability. Reinforcing this, J is considerably lower for the co-sputtered film even with a higher voltage applied.

To integrate these dielectrics in IGZO TFTs, devices with a staggered bottom gate, top-contact configuration were produced. Several devices were fabricated to analyze the effect of dielectric thickness, substrate bias (with or without substrate bias denoted SB and NSB, respectively), and annealing temperature (150 or 200 °C).

The output curve for a TFT using TSiO as dielectric layer, 200 nm thick, deposited by sputtering applying substrate bias and annealed at 200 °C, is shown in Fig. 3.8. It is clear that the operation regimes are well-defined, with the transistor operating in saturation for V$_{DS}$ >8 V for any V$_{GS}$. Moreover, in this regime, the curves present a good flatness indicating that the channel is fully depleted; increasing separation between curves as V$_{GS}$ increases suggests that there is no degradation of mobility for higher gate fields.

The evaluation of contact resistance is also imperative, as this has to be taken into account when designing the circuits. As it is possible to observe in

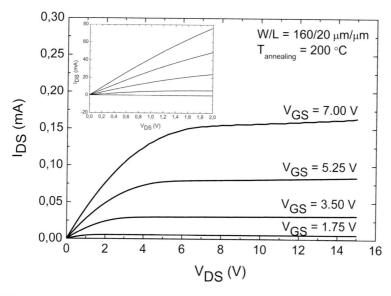

Fig. 3.8 Output characteristics of IGZO TFTs with Mo electrodes, using TSiO as dielectric layer ($\approx$200 nm), annealed at 200 °C. Inset shows magnification of linear regime

the inset of Fig. 3.8, there are no current crowding effects at low V_{DS} and the amplitude of the I_{DS} is considerably high. In fact, taking into account the work functions of Mo and IGZO, 4.7 and 4.5 eV, respectively, the Schottky barrier can be negligible, which means that this contact has a high efficiency of injection and, as a consequence, good electrical properties are obtained [17]. This is the case even for low annealing temperature (150 °C). Complete analysis of contact resistance would require measuring devices with different channel lengths and applying the transmission line method (TLM), which was not done in the framework of this work. Still, for similar devices in our laboratory TLM was applied being concluded that contact resistance only starts to be relevant for devices with channel length below 2 μm.

Figure 3.9 and Table 3.5 show the difference in TFT performance for devices with Ta$_2$O$_5$ and TSiO, using different dielectric thicknesses, with and without substrate bias during deposition, annealed at 200 °C. For TFTs with Ta$_2$O$_5$ dielectric a large increase of off-current (which also detrimentally affects S) is verified for lower dielectric thickness and NSB condition, proving the importance of these parameters to achieve good performance devices. Nevertheless, note that regardless of the processing conditions, Ta$_2$O$_5$-based transistors always present a large variation of properties over a substrate. In general, TFTs with TSiO present enhanced properties, particularly a closer to 0 V V_{on} and improved uniformity (in terms of device-to-device variation in the same substrate). As for Ta$_2$O$_5$, larger thickness and use of SB are advantageous for transistor performance for TSiO. In fact, when SB is used during a deposition, the molecules that are linked by weak bonds are

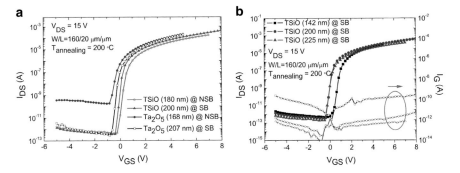

Fig. 3.9 Measured I–V characteristics for devices with W/L =160/20 μm/μm and annealed at 200 °C. Effect of dielectric composition and the usage of substrate bias in devices with approximately the same thickness (**a**) and of dielectrics thicknesses considering the same material and the usage of substrate bias (**b**)

Table 3.5 Summary of electrical properties of devices annealed at 200 °C using Ta_2O_5 and TSiO dielectrics processed under different conditions, as depicted in Fig. 3.9

Sample	Condition	Thickness (nm)	μ_{sat} (cm^2/V s)	On–off ratio	V_{on} (V)	S (V/dec)	I_G @ V_{Gmax} (A)
Ta_2O_5	SB	206.7	34.8[a]	6.62×10^8	−0.83	0.106	5.94×10^{-12}
Ta_2O_5	NSB	168.0	42.5[a]	1.21×10^6	−0.83	0.194	1.43×10^{-11}
TSiO	SB	142.0	30.1[a]	2.59×10^8	−0.21	0.123	1.55×10^{-12}
TSiO	SB	225.0	16.9	8.16×10^8	−0.50	0.128	7.00×10^{-13}
TSiO	NSB	180.2	20.1	3.80×10^8	−0.25	0.133	6.92×10^{-12}

[a] Given the large size of capacitors available in the mask layouts used in this work, Ta_2O_5 MIM structures typically exhibited large leakage currents, inhibiting a correct determination of C_i. Hence, μ_{sat} for TFTs with Ta_2O_5 and TSiO should be overestimated

re-sputtered, decreasing the deposition rates, as mentioned before. As a result, the molecules linked by strong bonds remain in film turning it denser, providing better insulating properties [5]. This argument is more evident in terms of S, suggesting that when SB is not used the interface dielectric/semiconductor presents worst properties.

In terms of I_G, lower values are obtained for thicker dielectrics, as expected (Fig. 3.9b). Furthermore, Table 3.5 also shows that the multicomponent approach and the use of substrate bias are able to reduce I_G levels when compared to Ta_2O_5. This indicates that good device reliability can be obtained by using thinner TSiO films, provided that substrate bias is used.

Another important consideration about TFT performance is related to hysteresis. Figure 3.10 shows this effect for Ta_2O_5 and TSiO and it is possible to conclude that hysteresis magnitude is smaller for TSiO, reinforcing the improved properties of this multicomponent dielectric over Ta_2O_5. Nevertheless, for both cases, counter-clockwise hysteresis is verified, suggesting ionic drift inside the dielectric layers. This is in fact supported by the positive gate-bias stress measurements performed on

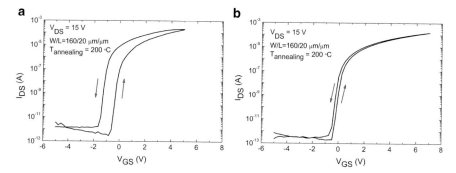

Fig. 3.10 Transfer curves in double sweep mode using Ta$_2$O$_5$ (**a**) and TSiO (**b**) with a thickness close to 200 nm and annealed at 200 °C. Both dielectrics were deposited by sputtering with substrate bias

devices annealed at 200 °C. Results for these experiments are presented in Fig. 3.11 for TFTs with both dielectrics (200 nm thick, deposited with substrate bias).

According to literature, there are two main mechanisms of instability that can occur in these TFT structures. The first one is related to electron trapping at or near the semiconductor/dielectric interface; the second is associated to the migration of negative ion within dielectric to gate/dielectric interface and the movement of positive ion within dielectric to semiconductor/dielectric interface [18]. Furthermore, the direction of V_{on} shift is imperative to understand the mechanism that is involved. In fact, the V_{on} shift is directly related to total charge that migrates to the new centroid location of charge. On the other hand, this shift is inversely proportional to gate capacitance. Some parameters such as light, room temperature, humidity, and the air exposition can also change the mechanism and its intensity.

Figure 3.11 suggests that the degradation mechanism involved with both dielectrics is associated with ion migration, as a negative ΔV_{on} is verified. This is plausible given the low deposition/annealing temperature of the dielectric layers used here, being the effect more intense for Ta$_2$O$_5$, in agreement with the hysteresis analysis presented above. In terms of recovery, this "instability" mechanism is reversible—the final V_{on} and S values are close to the last results obtained without supplying additional energy to the system. Hence, it is plausible to assume that defect creation should not be relevant, as it would typically require a subsequent annealing treatment to enable full recovery [18]. Still, TSiO restores its initial V_T faster than Ta$_2$O$_5$, reinforcing the idea that it is the best choice between these two dielectrics.

In brief, it was verified that multicomponent dielectrics, specially using substrate bias during deposition, provide better properties for dielectric layer. In fact, despite the relatively large variation of device properties during stress measurements, TSiO using substrate bias provides major advantages over single Ta$_2$O$_5$ layers deposited without substrate bias. Optimized TSiO devices present S in the range of 0.1–0.2 V/dec, on–off ratios exceeding 10^8 and $\mu_{sat} > 15 \text{ cm}^2/\text{V s}$. These results are comparable to state-of-the-art IGZO TFTs with the added advantages of

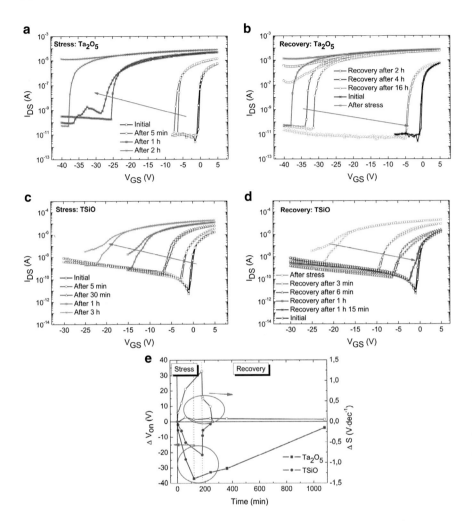

Fig. 3.11 Transfer characteristics evolution for stress and recovery for devices using single layers of Ta_2O_5, (**a**) and (**b**), and for TSiO (**c**) and (**d**). Summary of V_{on} and S variations during stress and recovery measurements, where *solid* and *open circles* denote ΔV_{on} and ΔS, respectively, (**e**). A positive bias stress was performed in air, dark, with a gate field of 0.16 MV/cm in TFTs with W/L = 160/20 $\mu m/\mu m$ and annealed at 200 °C

low operating voltage (enabled by the high-κ), low temperature processing, and having the same technique to deposit all the transistor layers. Furthermore, the smooth surface and amorphous structure make this material an excellent choice for integration into a multilayer dielectric configuration, which is the topic explored in the next section.

3.2.2   Multilayer Structures Based on TSiO and SiO₂

Although the results of one layer are satisfactory, it is suggested in the literature that multilayer structures can solve instability, off-current, and hysteresis problems when this configuration is integrated in devices [5, 19, 20].

To verify the improvements of multilayer configuration, two multilayers structures were considered, one with three layers (SiO_2–TSiO–SiO_2) and other using seven layers (SiO_2–TSiO–SiO_2–TSiO-SiO_2–TSiO–SiO_2), considering as reference the single layer TSiO film previously studied. In all samples, the expected thickness for the capping SiO_2 layers was approximately 20 nm and, for the intermediate layers of SiO_2, 10 nm. These thin films were produced using substrate bias.

Two multilayer samples were analyzed by RBS, one of them with three layers and expected thickness of 150 nm, the other with five layers[4] and estimated thickness of 250 nm. RBS conditions used during measurements were the same as for TSiO thin film. The thicknesses and Ta_2O_5/SiO_2 ratios for different samples analyzed by RBS are shown in Fig. 3.12 and Table 3.6.

RBS has more sensitivity for heavier elements than for lighter elements, especially if they are on lighter substrates. For this reason, the oxygen percentage included in different layers is not included in the table—a larger error is associated with it. Some differences are obvious between the spectra of the samples. Firstly, it is clear that the thickness in TSiO is higher because O and Ta peaks are broader compared to multilayer structures. Contrarily, the Si peak from multilayers is more evident which makes sense because these structures include layers with just SiO_2 while, in TSiO film, two materials were mixed in a single layer. Regarding the Ta peak in Fig. 3.12 (multilayer with five layers), it presents two overlapping peaks suggesting two layers with this material which is in accordance with a five layer sample (SiO_2–TSiO–SiO_2–TSiO–SiO_2). In terms of the Ta_2O_5/SiO_2 ratio, there are some differences between single and multilayers, which might be attributed to the different Ta_2O_5 targets used to sputter them, with the one used for multilayers providing a higher growth rate. This situation is common when using targets from different vendors and requires process adjustment. A more accurate fitting can also improve results and these discrepancies.

Multilayers structures with three and seven layers were also analyzed by spectroscopic ellipsometry in the same conditions used for single layer samples. The Tauc-Lorentz dispersion formula was used as well. Figure 3.13 shows the relations n-E and k-E for multilayers structures. The previous results of ellipsometry for single layers films are also represented in order to simplify the comparison. By fitting it was possible to obtain thicknesses, roughness, and E_G for the different dielectrics, shown in Table 3.7.

The TSiO layer was produced using similar conditions for single and multilayer configurations, which means that it is expected that these films have a similar n

---

[4]Seven layer structures were still not available when RBS analysis was performed but the five layer sample is perfectly suitable for the RBS comparisons envisaged here.

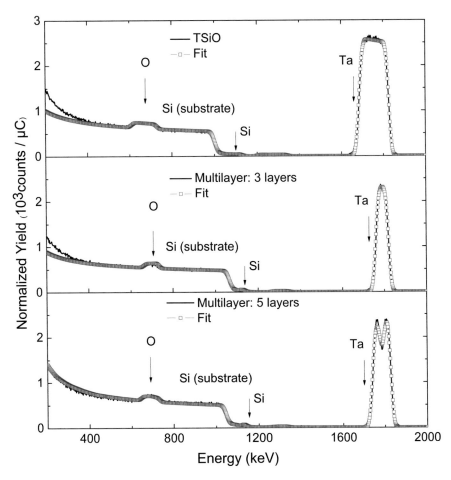

Fig. 3.12 Compositional analysis: comparison between TSiO and multilayers using three and five layers

Table 3.6 Parameters obtained for dielectrics after fitting for RBS results

Sample	Layer	Thickness (10^{15} at/cm^2)	Ta$_2$O$_5$/SiO$_2$ ratio
TSiO	–	1400	2.2
Multilayer: three layers	SiO$_2$	140	–
	TSiO	607	3.0
	SiO$_2$	123	–
Multilayer: five layers	SiO$_2$	91	–
	TSiO	368	3.0
	SiO$_2$	97	–
	TSiO	432	2.7
	SiO$_2$	117	–

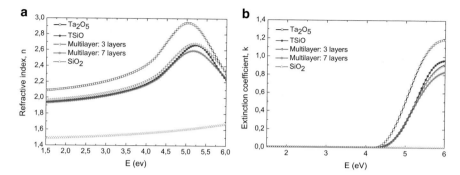

Fig. 3.13 Spectroscopic ellipsometry analysis: comparison between single and multilayers dielectrics thin films in terms of refractive index (**a**) and extinction coefficient (**b**) with SiO_2, as reference

Table 3.7 Properties of single and multilayers films obtained by Tauc-Lorentz dispersion method

Sample	Total thickness (nm)	Roughness (nm)	E_G (eV)
Ta_2O_5	267.4	1.1	4.20
TSiO	237.7	0.4	4.23
Multilayer: three layers	233.5 (17.1/190.2/26.2)	3.9	4.31
Multilayer: seven layers	248.5 (14.1/63.1/11.8/60.3/10.9/62.4/25.2)	10.1	4.32

Table 3.8 Electrical properties of dielectrics determined from MIM structures

Sample	Thickness (nm)	E_B (MV/cm)	κ	C_i (nF/cm^2)	J @ $E_{B\,ax}$(A/cm^2)
Ta_2O_5	230.0	1.54	22.6	87.0	1.21×10^{-2}
TSiO	210.0	2.81	17.0	71.7	8.88×10^{-2}
Multilayer: three layers	210.0	7.24	11.7	49.5	4.90×10^{-5}
Multilayer: seven layers	240.0	>7.5[a]	10.7	39.3	–

[a] Electrical breakdown not observed with the maximum voltage allowed with the Keithley 4200SCS

and k, and specially E_G. The small variations between referred films, in terms of n, are due to the difference in total thickness of TSiO film in structure, which influences the density of dielectric. Regarding roughness, the values obtained for multilayer structures should have some error associated because for the SiO_2/air interface the program considered as voids part of the SiO_2 film, which is supposed to have 20 nm and, for three and seven layers structures, the thicknesses obtained are smaller, 17.1 and 14.1 nm, respectively. Proceeding in the same manner as for the single layer study, MIM structures annealed at 150 °C were analyzed in order to extracted electrical parameters for these multilayer films. The obtained results are summarized in Table 3.8.

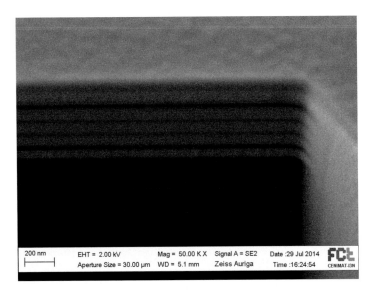

Fig. 3.14 Cross section of TFT near the channel (SEM image)

Regarding κ, the values are in accordance with previous analysis, wherein it is evidently a decrease for films with a poor Ta concentration. The smallest result for κ is close to 10, which is still significantly larger than the 3.9 value of SiO_2. Simultaneously, the E_B and J verified for the stack with seven layers are also excellent, clearly showing that the multilayer approach is effective to significantly improve the insulating performance of the dielectric.

Using SEM-FIB it was possible to prepare and image a cross section near the channel in order to identify all layers present in the device (Fig. 3.14). No significant effects from lithographic processes or layer heterogeneities were detected. Starting from the bottom upwards, layers could correctly be identified as follows:

- Substrate (thick, Dark);
- Mo gate electrode ($\approx$60 nm);
- Multilayer dielectric:
    SiO_2 ($\approx$20 nm);
    TSiO ($\approx$60 nm);
    SiO_2 ($\approx$10 nm);
    TSiO ($\approx$60 nm);
    SiO_2 ($\approx$10 nm);
    TSiO ($\approx$60 nm);
    SiO_2 ($\approx$20 nm);
- IGZO—semiconductor ($\approx$30 nm);
- Mo—source–drain electrode ($\approx$60 nm);

For most of the devices annealed at 200 °C fabricated during this work, a hump is verified in the transfer characteristics, presumably related with the activation of

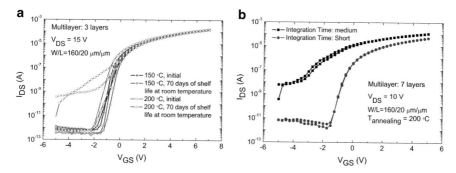

Fig. 3.15 Transfer characteristics for devices annealed at different temperature and evaluated in different times, using a medium integration time (**a**). Effect of integration time in I–V characteristics (**b**)

some contamination arising from processing and/or oxygen ion migration during annealing, increasing trap density. Still, the nature of these traps is not entirely clear so far and no traces of contaminations could be found within the detection limits of the EDS detector [21]. After a stabilization period, i.e., air exposure during 70 days (due to their staggered bottom gate configuration, the semiconductor is directly in contact with environment), a "rebalancing" of these molecules occurs and the hump is considerably attenuated (Fig. 3.15a). After this stabilization period it is found that the annealing temperature does not induce a significant variation in TFT performance. As a consequence, the lowest temperature (150 °C) is preferable, as it does not present this hump effect and assures compatibility with flexible and paper electronics. Another interesting result is found when comparing transfer characteristics taken with medium and short integration times (essentially, changing the time taken by the semiconductor analyzer to collect each datapoint), as presented in Fig. 3.15b. Even if equipment resolution is naturally degraded when short integration time is used, lower I_G (hence, off-current) is recorded and less non-idealities are found in the transfer curve, providing hints that fast states should be the main causes for such non-idealities (although the nature of these states is not known at this stage and certainly deserves further investigation in the future).

For three layers configuration, three different thicknesses were evaluated, ranging from 100 to 250 nm. Taking into account that the TFT with 200 nm of dielectric layer provided the best performance, the seven layers structures were produced considering approximately the same thickness. All the extracted electrical parameters are presented in Table 3.9.

In terms of stress, for both multilayer structures, a negative and then a positive V_{on} shift is verified, indicating that two instability mechanisms are involved (Fig. 3.16e). Firstly, for multilayer with three layers, a negative shift occurs during the first 30 min stress time, but after that period V_{on} starts shifting to positive voltages. For multilayer with seven layers, this change of V_{on} direction happens after only 5 min. This variation between negative and positive shifts is indicative that the ion migration is the first mechanism activated but, after that, the electron

Table 3.9 Electrical properties for TFT devices that used multilayers structures (W/L = 160/20 μm/μm)

Multilayer	Temperature (°C)	Thickness (nm)	μ_{sat} (cm²/V s)	On–off ratio	V_{on} (V)	S (V/dec)	I_G @ $V_{G\,max}$ (A)
Three layers	150	250.0	14.5	4.21×10^8	−0.75	0.176	1.51×10^{-10}
Three layers	150	200.0	14.6	8.36×10^8	−2.00	0.177	6.80×10^{-13}
Three layers	150	100.0	24.9	6.31×10^5	−0.25	0.182	1.78×10^{-10}
Seven layers	200	240.0	12.7	3.23×10^7	−1.50	0.213	1.26×10^{-10}

Devices were measured using a short integration time to prevent hump effects, which increases the measured I_G values over more conventional medium integration times

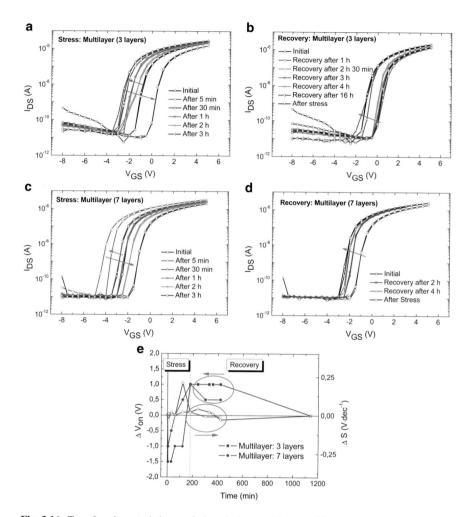

Fig. 3.16 Transfer characteristics evolution during positive gate-bias stress and recovery, for IGZO TFTs with multilayer dielectrics: three layers (**a**) and (**b**); seven layers (**c**) and (**d**). Summary of V_{on} and S variations during stress and recovery measurements, where *solid* and *open circles* denote ΔV_{on} and ΔS, respectively, (**e**). A positive bias stress was performed in air, dark, with a gate field of 0.16 MV/cm in TFTs with W/L = 160/20 μm/μm and annealed at 200 °C

trapping starts to be the dominant mechanism. These V_{on} shifts are not accompanied by a significant variation of S, showing that new defect states are not created. The evolution of transfer characteristics during stress and recovery measurements is shown in Fig. 3.16. Large improvement is verified when comparing these plots with the ones of TFTs having single layer dielectrics (Fig. 3.11).

In brief, taking into account all results obtained, it is clear to conclude that the multilayer structures greatly improve the TFT performance and stability and are preferable to include in circuits with a high level of complexity. In fact, the seven layer dielectrics exhibit a large E_B and low I_G, while maintaining a reasonably high-κ that allows to induce large charge densities in the semiconductor and assure low voltage operation.

As a final note, having this work as a reference, more improvements were done in the last months to the seven layer dielectric, tuning the thickness of individual layers as well as of the full stack. IGZO TFTs annealed at only 150 °C using this improved multilayer dielectric, whose thickness is only 115 nm, exhibit $\mu_{sat} \approx 20\,cm^2/V\,s$, on–off ratio $> 10^8$, $V_{on} = -1.5\,V$, S=0.16 V/dec and $I_G \approx 1\,pA$. Results on this will be published soon.

3.3 IGZO TFT Modeling

To successfully design and produce either basic or complex circuits, the use of an accurate device model is imperative and should take into account all the aspects that influence the device performance.

Given the very early stage of oxide TFTs development, specific models for IGZO devices are not yet widespread. However, some efforts have been done in this field where some models have been created or adjusted based on Verilog A, neural networks or Spice [22, 23].

In this work, an existent a-Si TFT model developed by Semiconductor Devices Research Group at RPI is adjusted to IGZO TFTs (Fig. 3.17). In this model, the dependence between μ_{FE} and V_{GS} is considered. This is highly relevant in the framework of oxide TFTs, where it is well known that as V_{GS} increases Fermi level can penetrate into the conduction band and greatly enhance carrier transport. Capacitance per unit area, based on MIM structures characterization, was included in the model. However, C_{GS}, C_{GD}, and C_{GG} characterizations should be performed in a future work in order to build an accurate AC and transient model.

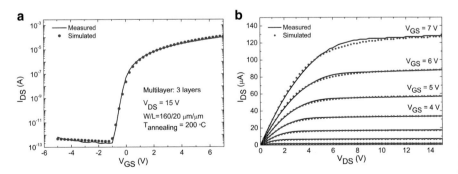

Fig. 3.17 Measured and simulated I–V characteristics for a multilayer device (3 layers) with W/L=160/20 μm/μm: transfer curve (**a**) and output curve (**b**)

References

1. P. Barquinha, Transparent oxide thin-film transistors: production, characterization and integration. Ph.D. thesis, 2010
2. H.Q. Chiang, Development of oxide semiconductors: materials, devices, and integration. Ph.D. thesis, Oregon State University, 2007
3. A.H. Simon, Sputter processing, in *Handbook of Thin Film Deposition*, 3rd edn., ed. by K. Seshan (William Andrew Publishing, Oxford, 2012), pp. 55–88
4. R.C. Jaeger, *Introduction to Microelectronic Fabrication* (Addison-Wesley Longman Publishing, Boston, 1987)
5. P. Barquinha, R. Martins, L. Pereira, E. Fortunato, *Transparent Oxide Electronics: From Materials to Devices* (Wiley, Chichester, 2012)
6. D. Hess, Dry-etching processes, in *Microelectronic Materials and Processes*, ed. by R.A. Levy (Springer Netherlands, Dordrecht, 1989)
7. K. Nojiri, *Dry Etching Technology for Semiconductors* (Springer International Publishing, Cham, 2015)
8. H. Stanjek, W. Häusler, Basics of X-ray diffraction. Hyperfine Interact. **154**(1–4), 107–119 (2004)
9. C.R. Blanchard, Atomic force microscopy. Chem. Educ. **1**(5), 1–8 (1996)
10. G. Binnig, C.F. Quate, C. Gerber, Atomic force microscope. Phys. Rev. Lett. **56**(9), 930–933 (1986)
11. Y. Leng, *Materials Characterization* (Wiley-VCH Verlag GmbH & Co. KGaA, Weinheim, 2013)
12. N.P. Barradas, C. Jeynes, R.P. Webb, Simulated annealing analysis of Rutherford backscattering data. Appl. Phys. Lett. **71**(2), 291 (1997)
13. W.-K. Chu, J.W. Mayer, M.-A. Nicolet, *Backscattering Spectrometry* (Academic, New York, 1978)
14. H. Fujiwara, *Spectroscopic Ellipsometry* (Wiley, Chichester, 2007)
15. D.K. Schroder, *Semiconductor Material and Device Characterization* (Wiley, Hoboken, 2005)
16. C. Chaneliere, S. Four, J. Autran, R. Devine, Comparison between the properties of amorphous and crystalline Ta2O5 thin films deposited on Si. Microelectron. Reliab. **39**(2), 261–268 (1999)
17. P. Barquinha, A.M. Vila, G. Gonçalves, L. Pereira, R. Martins, J.R. Morante, E. Fortunato, Gallium-indium-zinc-oxide-based thin-film transistors: influence of the source/drain material. IEEE Trans. Electron Devices **55**(4), 954–960 (2008)
18. J.F. Wager, Transparent electronics. Science **300**(5623), 1245–1246 (2003)
19. L. Zhang, J. Li, X.W. Zhang, X.Y. Jiang, Z.L. Zhang, High-performance ZnO thin film transistors with sputtering SiO2/Ta2O5/SiO2 multilayer gate dielectric. Thin Solid Films **518**(21), 6130–6133 (2010)
20. L. Zhang, H. Zhang, J.W. Ma, X.W. Zhang, X.Y. Jiang, Z.L. Zhang, Copper phthalocyanine thin-film field-effect transistor with SiO2/Ta2O5/SiO2 multilayer insulator. Thin Solid Films **518**(21), 6134–6136 (2010)
21. D. Kang, H. Lim, C. Kim, I. Song, J. Park, Y. Park, J. Chung, Amorphous gallium indium zinc oxide thin film transistors: sensitive to oxygen molecules. Appl. Phys. Lett. **90**(19), 192101 (2007)
22. G. Bahubalindruni, V.G. Tavares, P. Barquinha, C. Duarte, R. Martins, E. Fortunato, P.G. de Oliveira, Basic analog circuits with a-GIZO thin-film transistors: modeling and simulation, in *2012 International Conference on Synthesis, Modeling, Analysis and Simulation Methods and Applications to Circuit Design (SMACD)* (IEEE, New York, 2012), pp. 261–264
23. D.H. Kim, Y.W. Jeon, S. Kim, Y. Kim, Y.S. Yu, D.M. Kim, H.-I. Kwon, Physical parameter-based spice models for InGaZnO thin-film transistors applicable to process optimization and robust circuit design. IEEE Electron Device Lett. **33**(1), 59–61 (2012)

Chapter 4
Analog-to-Digital Converters

Abstract Nowadays, data converters are considered one of the most relevant blocks in electronics, having a large amount of applications. In the last years, a fast trend in different fields, such as in digital processing and in semiconductor industry, boosted the performance of analog-to-digital converters (ADCs) and of digital-to-analog converters (DACs).

In this chapter a concise historical perspective and a brief background related to ADCs is given. The main architectures are referred, emphasizing the operation mode of successive approximation ADCs (SAR-ADCs) and of sigma-delta ($\Sigma\Delta$) ADCs, taking into account their current relevance. Additionally, and given the purpose of this book, the usage of thin film technologies in circuits is analyzed, especially when they are applied in ADCs.

4.1 ADCs: Relevance, Historical Perspective, and Main Architectures

The fast trend towards digital processing of analog signals in an increased number of application fields has stimulated significant research efforts in the area of data converters implemented in many different types of technologies. Analog-to-digital converters (ADCs) are, therefore, paramount in modern systems where increasingly complex processing of analog signals is performed digitally.[1]

The ADCs appearance is clearly related with semiconductor industry, which influenced the market trends since 1960. At that stage, the application of semi-conductors was extremely intensified. According to Gartner, Inc., the investment in semiconductor industry grows from $1 billion in 1964 to $348 billion in 2015, influencing directly the integrated circuits (ICs) market [1]. In fact, the growth of semiconductor industry has been so forceful that new technologies have been developed. Despite CMOS node has been decreasing, improving speed and

---

[1]The opposite function, conversion from a digital signal into an analog signal is also extremely relevant in this field. However, digital-to-analog converters (DACs) will not be explored in this book.

© The Author(s) 2016   49
A.P.P. Correia et al., *A Second-Order ΣΔ ADC Using Sputtered IGZO TFTs*,
SpringerBriefs in Electrical and Computer Engineering,
DOI 10.1007/978-3-319-27192-7_4

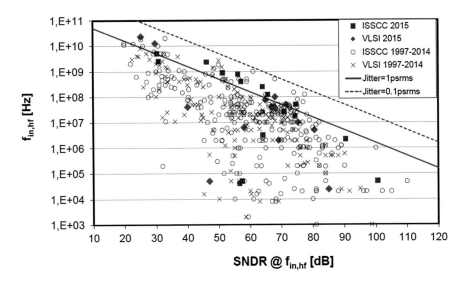

Fig. 4.1 Relevant ADCs publications in ISSCC and in VLSI between 1997 and 2015 taking into account their SNDR and F_{in} [2]

die area of chips, other problems related with parasitic capacitances, modelling and leakage current started to be relevant and compromise the performance of circuits. Consequently, solutions such as high-κ dielectrics, metal gate, or even new approaches [e.g., the usage of thin-film transistors (TFTs)] begin to be implemented not only in the scope of academic research but also in industry. Consequently, techniques and architectures used in ADCs suffered some relevant improvements, in order to promote lower operating voltage, higher sampling rate (F_S), higher input frequency (F_{in}), and resolution (N).

Figure 4.1 shows a wide range of quality publications in the IEEE International Solid-State Circuits Conference (ISSCC) and in the Very Large Scale Integration (VLSI) Symposia between 1997 and 2015 [2]. Reinforcing the fast growth trend, a significant improvement in signal-to-noise distortion ratio (SNDR) is verified even for higher F_{in}.

In terms of operation, an ADC implements the function of an analog divider (since the analog input is divided by a constant reference), by converting a continuous-time and continuous-amplitude analog signal (V_{in}) into a digital signal (D_{out}), i.e., discrete in both time and amplitude [3].

Basically, there are two important steps in analog-to-digital (A/D) interfaces—sample-and-hold (S/H) and quantization, shown in Fig. 4.2. The first one is responsible to sample an analog signal, at a given rate, and hold it (time discretization). Secondly, in the quantization step, a quantizer circuit determines the discrete value that represents the amplitude of the input signal. After that, a digital code is provided at the output by an encoder digital block.

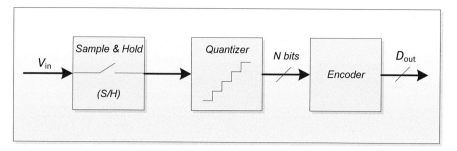

Fig. 4.2 Generic block diagram of an ADC

In addition, ADCs can be distinguished between Nyquist-rate and oversampling A/D converters depending on relation between input and output signals. For the first family of A/D converters, the relation is one-to-one and the Nyquist theorem is fulfilled. For oversampling A/D converters, F_S is higher than the necessary by the Nyquist criterion and an oversampling ratio (OSR) is defined by $F_S/2 \cdot$ BW, where BW is the signal bandwidth. Some characteristics of ADCs are intensely relevant to evaluate their performance. Between them, it is proper to consider F_S, N, SNDR, and power dissipation [4]. However, their relative importance is highly dependent on a given application.

Regarding A/D architectures, there are a considerable number of them such as parallel flash, successive approximation ADCs (SAR-ADCs), $\Sigma\Delta$, pipeline and, among many others, which can involve the hybrid combination between different architectures. The proper choice depends on the application requirements (N, BW, F_S, etc.). According to the Murmann's database [2], there are two architectures that present an intense grown in this last years: SAR-ADCs and $\Sigma\Delta$. SAR-ADCs provide a moderate speed and a reasonable level of circuit complexity, while $\Sigma\Delta$ modulators ($\Sigma\Delta$Ms) have been largely studied due to their attractive characteristics in terms of high resolution and of low voltage operation which are very important for high-quality applications such as audio or biomedical applications [5]. Given the importance of these architectures, SAR-ADCs and $\Sigma\Delta$ ADCs will be briefly explained.

4.1.1   Successive Approximation ADCs (SAR-ADCs)

SAR-ADCs have been proposed, in recent years, as one of the most popular approaches for realizing A/D conversion due to their moderate speed and low circuit complexity. The first description of the SAR-ADC architecture is reported back to 1947, and it can be found in a journal paper, by W.M. Goodall, from Bell Labs. Nowadays, with SAR-ADCs it is possible to achieve moderate resolutions (10–14 bits) with the absence of amplifiers.

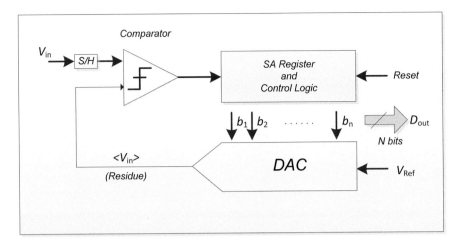

Fig. 4.3 Block-diagram of a Nyquist-rate SAR-ADC

Figure 4.3 shows the block diagram of a SAR-ADC. It basically comprises a front-end sample-and-hold (S/H), a comparator, a DAC, a successive approximation (SA) register (SAR), and some additional control logic (CL). A SAR-ADC performs a "binary search" to find the closest digital word, D_{out}, (corresponding to the closest analog estimate, ($< V_{in} >$)) to match the (sampled-and-held) analog input signal (V_{in}). The original SA algorithm was initially called "feedback subtraction" and it can be traced back to the sixteenth century, related with the solution for finding the determination of an unknown weight by a minimal sequence of weighting operations. Furthermore, from the mathematical point of view, the SAR-ADC and the Recycling-ADC architectures are equivalent. In general, a "binary search" algorithm divides the search space in two each time, and the desired digital output word can be found in N steps for a set of organized data of size 2^N. Specifically, after an initial reset and in the first clock-cycle, the most significant bit (MSB), b_1, is determined. In the second clock-cycle, the next bit, b_2, is determined, followed by b_3, and so on, until all N bits are found. The main goal of the "binary search" algorithm is to reduce the analog "residue" to less than one least significant bit (LSB). Therefore, in its most straightforward implementation, a SAR-ADC requires N clock-cycles to complete an N-bit conversion [3].

4.1.2 Sigma–Delta ($\Sigma\Delta$) Modulators

In $\Sigma\Delta M$ ADCs, there are two underlying concepts: oversampling and noise shaping. The ADCs that use oversampling do not need to have a rigorous matching tolerance, and another advantage is that it simplifies the requirements of the anti-aliasing filters. On the other hand, noise shaping is responsible to increase

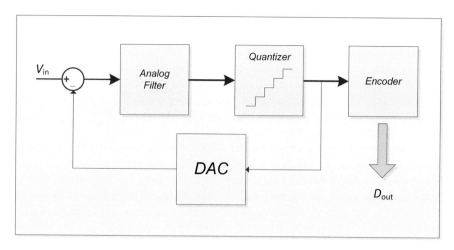

Fig. 4.4 Simplified schematic of an oversampling ADC with noise shaping

signal-to-noise ratio (SNR) because it alters ("shapes") the frequency distribution of the quantization noise—the noise density is higher at frequencies where is not so problematic (since the noise will be mainly concentrated at higher frequencies, i.e., out of the band of interest and, therefore, easily filtered by a digital decimation filter) [6].

As shown in Fig. 4.4, this structure is basically built using an analog filter, a quantizer (with N bits) and a DAC with a certain number of bits. Through the feedback-loop that this architecture makes use of, imprecisions are internally adjusted because the quantization error is subtracted from input signal [7].

The most critical blocks in the entire $\Sigma\Delta M$ are the analog filter (integrator(s)) and the quantizer. In the case of a single-bit quantizer ($N=1$), it simply relies on a single comparator [3]. Moreover, the number of integrators employed in the circuit mainly defines the order of the analog filter. Comparing with first-order $\Sigma\Delta M$ ADCs, higher orders allow improving the SNR. However, orders higher than two can suffer from stability problems. Consequently, the second-order architecture is widely adopted. Another important feature of $\Sigma\Delta M$ ADCs is related to the type of implementation of the analog filter (the integrator(s)), which defines whether this modulator is operating continuously in time (CT) or in the discrete in time (DT) domain. In fact, CT $\Sigma\Delta M$ architectures have been largely studied because they allow achieving a larger BW when compared to their DT $\Sigma\Delta M$ counterparts.

4.2 ADCs Using Thin-Film Technologies

Large area electronics relies to a great extent on thin-film technologies. In particular, the evolution of the performance levels achieved with TFTs in the last years together with the advances in their processing technologies is enabling the design and fabrication of increasingly complex and fast electronic circuits. For semiconductor materials such as organics or oxides, owing to their low processing temperatures, circuits can even be made in flexible foils, roll-to-roll, and printing systems. These scientific and technologic advances definitely boost the range of applications possible for thin films circuitry, associating large area uniformity, high performance, and integration levels to characteristics such as transparency, recyclability, and low cost. In brief, embedding electronic functionalities in every kind of surfaces and shapes is starting to be a reality.

Despite the tremendous importance of a-Si:H, organic or low temperature poly-crystalline silicon (LTPS) TFTs in advancing thin-film devices and circuits over the last decades, oxide TFTs have a large potential to become a mainstream technology in the next few years. To the best author's best of knowledge, demonstrations of circuit integration started in 2006, with n-type MOS inverters and a five-stage ring oscillator based on indium-gallium oxide (IGO) TFTs with a maximum oscillation frequency of 9.5 kHz being reported [8]. Following this, several digital circuits were physically demonstrated with oxide TFT technology, such as shift registers and gate drivers [9, 10], mostly aiming to the integration of those in displays. As a demonstration of higher integration capability, DACs were reported with IGZO TFTs [11].

Some reports related with practical realizations of ADCs using TFTs are summarized in Table 4.1. Although being a small part of all developed work in this area, it contains examples of ADCs employing a-Si:H, poly-Si, LTPS, and organic TFTs. More details such as fabrication conditions or measurements results can be found in [12–16].

Concerning the implementation and development of ADCs, two possible decisions can be made:

- Search and research analog circuit techniques (e.g., the use of feedback) and architectures (sigma–delta modulation with high-order noise shaping and intensive use of passive circuitry) that cope with the inherent variability and performance limitations of TFT technologies (e.g., intrinsic gain of the devices, threshold voltage variability, etc.);
- Implement simple A/D architectures (e.g., the successive approximation A/D architecture) together with dedicated calibration algorithms that use digital circuitry to assist the poor analog technological performance.

The former option is the premise for the work developed and reported throughout this book.[2]

---

[2]It is extremely relevant to emphasize that, to the best of the authors' knowledge, and up to date no kind of reference was found regarding the use of oxide semiconductors TFTs in designing ($\Sigma\Delta$) ADCs.

Table 4.1 ADCs using thin-film technologies

Technology	Architecture	Year	ADC characteristics	Authors
a-Si:H (n-type)	Flash (2 kS/s)	2012	5 bits Without calibration $V_{DD} = 10$ V Power consumption: 13.6 mW	Dey and Allee [12]
Poly-Si TFTs (laser crystallized)	Flash (3 MS/s)	2010	3 bits Stainless steel foil substrate	Jamshidi-Roudbari et al. [13]
LTPS TFTs	Second-order $\Sigma\Delta$	2009	$V_{DD} = 11.2$ V DR=69 dB SNDR=65.63 dB Power consumption: 63.3 mW	Lin et al. [14]
Complementary organic TFTs	SAR ($F_S = 100$ Hz)	2010	6 Bits Power consumption: 3.6 W	Xiong et al. [15]
Organic TFTs	First-order $\Sigma\Delta$ (OSR=16)	2011	Flexible plastic foil SNR=26.5 dB $V_{DD} = 15$ V Power consumption: 1.5 mW	Marien et al. [16]

References

1. C. STAMFORD, Gartner says worldwide semiconductor sales expected to reach \$348 billion in 2015, a 2.2 percent increase from 2014, 2015 [Online]. Available http://www.gartner.com/newsroom/id/3089917
2. B. Murmann, Adc performance survey 1997–2015 [Online]. Available http://web.stanford.edu/~murmann/adcsurvey.html (visited on 1 Oct 2015)
3. T.C. Carusone, D. Johns, K. Martin, *Analog Integrated Circuit Design*, 2nd ed. (Wiley, 2011). http://eu.wiley.com/WileyCDA/WileyTitle/productCd-EHEP002039.html
4. B. Le, T. Rondeau, J. Reed, C. Bostian, Analog-to-digital converters. IEEE Signal Process. Mag. **22**(6), 69–77 (2005)
5. J.L.A. De Melo, A low power 1-MHz continuous-time $\Sigma\Delta$M using a passive loop filter designed with a genetic algorithm tool, in *Proceedings - IEEE International Symposium on Circuits and Systems*, vol. 1 (2013), pp. 586–589
6. E. Janssen, A. van Roermund, *Look-Ahead Based Sigma-Delta Modulation* (Springer Publishing Company, New York, 2013), p. 283
7. H. Marien, M. Steyaert, P. Heremans, *Analog Organic Electronics* (Springer, New York, 2013)
8. R. Presley, D. Hong, H. Chiang, C. Hung, R. Hoffman, J. Wager, Transparent ring oscillator based on indium gallium oxide thin-film transistors. Solid State Electron. **50**(3), 500–503 (2006)
9. M. Mativenga, M.H. Choi, J.W. Choi, J. Jang, Transparent flexible circuits based on amorphous-indium-gallium-zinc-oxide thin-film transistors. IEEE Electron Device Lett. **32**(2), 170–172 (2011)
10. B. Kim, C.-I. Ryoo, S.-J. Kim, J.-U. Bae, H.-S. Seo, C.-D. Kim, M.-K. Han, New depletion-mode IGZO TFT shift register. IEEE Electron Device Lett. **32**(2), 158–160 (2011)
11. D. Raiteri, F. Torricelli, K. Myny, M. Nag, B. Van der Putten, E. Smits, S. Steudel, K. Tempelaars, A. Tripathi, G. Gelinck, A. Van Roermund, E. Cantatore, A 6 b 10 MS/s current-steering DAC manufactured with amorphous gallium-indium-zinc-oxide TFTs achieving SFDR > 30 db up to 300 kHz, in *2012 IEEE International Solid-State Circuits Conference* (IEEE, New York, 2012), pp. 314–316
12. A. Dey, D.R. Allee, IEEE, Amorphous silicon 5 bit flash analog to digital converter, in *2012 IEEE Custom Integrated Circuits Conference*, 2012
13. A. Jamshidi-Roudbari, P.-C. Kuo, M.K. Hatalis, A flash analog to digital converter on stainless steel foil substrate. Solid State Electron. **54**(4), 410–416 (2010)
14. W.-M. Lin, C.-F. Lin, S.-I. Liu, A CBSC second-order sigma-delta modulator in 3 μm LTPS-TFT technology, in *2009 IEEE Asian Solid-State Circuits Conference* (IEEE, New York, 2009), pp. 133–136
15. W. Xiong, U. Zschieschang, H. Klauk, B. Murmann, A 3v 6b successive-approximation ADC using complementary organic thin-film transistors on glass, in *2010 IEEE International Solid-State Circuits Conference - (ISSCC)* (IEEE, New York, 2010), pp. 134–135
16. H. Marien, M.S.J. Steyaert, E. van Veenendaal, P. Heremans, A fully integrated $\Delta\Sigma$ ADC in organic thin-film transistor technology on flexible plastic foil. IEEE J. Solid State Circuits **46**(1), 276–284 (2011)

Chapter 5
A Second-Order ΣΔ ADC with Oxide TFTs @ FCT-UNL

Abstract The capability of new technologies, such as oxide thin-film transistors (TFTs) is reinforced when they are successfully integrated in circuits or systems with a considerable level of complexity. In the particular case of this study, sputtered indium-gallium-zinc oxide (IGZO) TFTs are used to design, simulate and produce an analog-to-digital converter (ADC), demonstrating the potential of this technology namely in terms of lower processing temperature and costs and excellent performance, typically far superior to a-Si and organics and comparable to LTPS.

In this chapter, the optimized IGZO TFTs with multilayer and multicomponent dielectric, based on Ta_2O_5 and SiO_2, are used to design a sigma-delta ($\Sigma\Delta$) modulator ($\Sigma\Delta$M), with a particular attention to the comparator, the active block in the circuit. The circuits are simulated using commercial electronic design automation (EDA) tools, which will be briefly referred. Furthermore, some layout considerations and hints for a successful fabrication are also provided.

5.1 Simulation and EDA Tools

EDA tools acquire an important role for designing electronic circuits, such as integrated circuits (ICs). Hence, it is possible to analyze and forecast the device operation and, consequently, avoid either designing or realizing erroneous circuits. For designs employing more than a few transistors, built under different technologies, EDA tools are crucial. Cadence Design Systems, Inc. is a well-known company that offers some relevant tools, supporting design, simulation, layout, and complete verification of the circuits to be submitted for fabrication.

In This Work The circuit design and its simulation was realized using Virtuoso[TM] Platform and Spectre[TM] Simulator from CADENCE. Despite the circuit optimization from an electrical point of view, the layout was also executed using this tool which allows the files extraction for subsequent mask fabrication. This last stage involved the development of a dedicated parameterized cell (PCELL) using design rules based on the existing fabrication processes available at CENIMAT Lab.

© The Author(s) 2016
A.P.P. Correia et al., *A Second-Order ΣΔ ADC Using Sputtered IGZO TFTs*,
SpringerBriefs in Electrical and Computer Engineering,
DOI 10.1007/978-3-319-27192-7_5

5.2   Comparator: Circuit and Simulation Results

Comparators are widely used in electronics and they play a relevant role in ADCs, being one of the most important building block in ADCs' architectures. Consequently, there are a considerable number of approaches depending on applications and the required specifications for the block [1].

Given the very early stage of oxide TFTs development, where IGZO TFTs are included, some important assumptions were done at the beginning of this work involving all blocks:

- Due to the lack of a reproducible and stable p-type oxide TFTs, blocks have been designed using only n-type devices;
- Diode-connected loads were preferred considering their improved mismatch over the sputtered passive elements available in-house;
- Targeting an improved fabrication yield, a minimum multiplicity of 2 has been adopted and used for the circuit design and optimization;
- Initially, comparator and sub-blocks were initially designed using two fixed aspect ratios (W/L) and sizes for TFTs (160/20 and 40/20 μm/μm) in order to simplify the design and layout. As consequence, the comparator results presented in this sub-section take into account this assumption. However, during $\Sigma\Delta$M design, other TFT sizes have been considered (e.g., channel lengths of 10 μm) in order to improve circuit speed (more details will be given in Sect. 5.3).

Concerning design and simulation, an inverter, considered as the simplest circuit, was firstly studied using IGZO TFTs model previously adjusted (Sect. 3.3). In this case, the circuit was designed to have a load device (n-type diode-connected) and a driver TFT. In order to adjust the output voltages for high and low levels to the maximum and the minimum, respectively, and taking into account that V_{DD} =10 V, the sizing ratio between driver and load was defined to 4:1. The simulation results of an inverter (considering devices with a channel length of 10 μm), shown in Fig. 5.1, indicate that the output voltage needs some μs in order to reach a well-defined logic value, which limits the operating frequency of circuits.[1]

The complete architecture of the used comparator is shown in Fig. 5.2. It encompasses a cascade of three pre-amplification (pre-amp) stages followed by a positive-feedback latch (PFBL) stage and, at the end of the chain, four logic inverters with feedback, implementing a fully dynamic digital latch (for output regeneration and memory). These blocks are respectively depicted in Fig. 5.2a–c.

Regarding pre-amp block, each stage comprises a differential pair ($M_{2a,b}$) driving n-type diode-connected loads ($M_{1a,b}$). The PFBL uses two n-type analog inverters (devices $M_{3a,b}$ and $M_{5a,b}$) cascaded with two analog latches (cross-coupled transistor-pairs, $M_{4a,b}$ and $M_{6a,b}$) to improve the comparison speed [2]. In addition,

---

[1]In this simulation, the length of devices was sized to 10 μm since this image was obtained during re-design of comparator, to be integrated in $\Sigma\Delta$M.

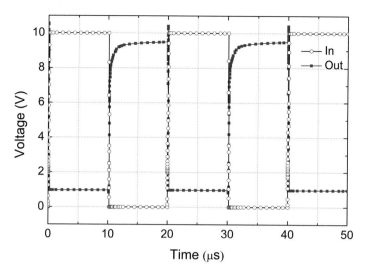

Fig. 5.1 Simulation result of an inverter considering the load n-type TFT (40/10 μm/μm) and driver (160/10 μm/μm)

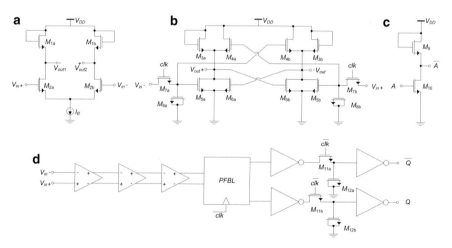

Fig. 5.2 Schematic of the proposed comparator: (**a**) pre-amplifier (pre-amp); (**b**) positive-feedback latch (PFBL); (**c**) logic inverter; (**d**) complete comparator

to increase the regeneration-speed, devices M_{4a} and M_{4b} are cross-coupled to the inputs of the n-type analog inverters.

In terms of operation, when the clock signal (*clk*) is disabled, the differential input is amplified by the cascade of the three pre-amp stages, with enough DC gain to overcome a possible large offset in the PFBL stage. The supply current in the PFBL is cut off and the outputs (before the digital output inverters) hold the memory of the previous state. If a seed signal is applied differentially to devices $M_{7a,b}$ and

Table 5.1 Transistors' dimensions used in the initial sizing of the comparator

Transistor	Characteristics		
	Width (μm)	Length (μm)	Multiplicity
$M_{1a,b}$	40	20	2
$M_{2a,b}$	160	20	4
$M_{3a,b}$	40	20	2
$M_{4a,b}$	160	20	2
$M_{5a,b}$	40	20	4
$M_{6a,b}$	40	20	8
$M_{7a,b}$	40	20	8
$M_{8a,b}$	160	20	2
M_9	40	20	2
M_{10}	160	20	2
$M_{11a,b}$	40	20	2
$M_{12a,b}$	40	20	2

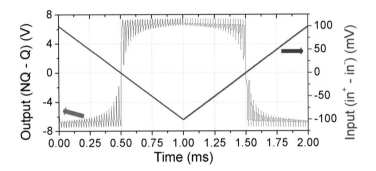

Fig. 5.3 Differential output when an input triangular signal is applied

the *clk* pulse goes high, the output of the stronger side of M_{5a} and M_{5b} is pulled down more strongly than the other side. This will cause the PFBL made of devices $M_{4a,b}$ and $M_{6a,b}$ to flip to one of the two stable states. While n-type capacitors $M_{8a,b}$ help adjusting the correct time-constant associated with the regenerative poles for proper operation, n-type capacitors $M_{12a,b}$ provide dynamic memory to the output digital latch.

It is noteworthy the sizing of circuit was done taking into account the simulation results (i.e., by trial-and-error) since, although feasible, to derive a complete set of analytical equations is not an easy task (mainly for the PFBL circuit due to its intrinsic positive-feedback nature).

Table 5.1 shows the transistors' dimensions used in the comparator. If higher accuracy is envisaged, the number of cascaded pre-amp stages can be increased as well as the multiplicity of devices $M_{2a,b}$. However, die area will increase and speed will be reduced.

The simulation setup was based on two major tests. Firstly, a fully differential input triangular signal with 100 mV amplitude (as shown in Fig. 5.3) has been

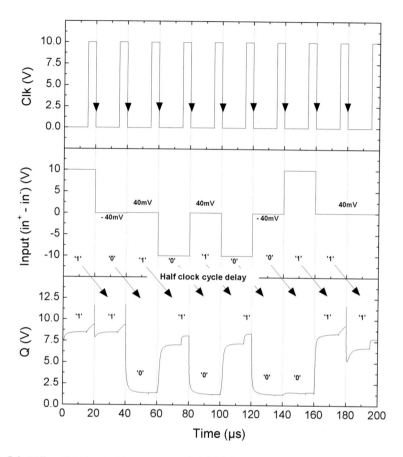

Fig. 5.4 Differential signal with a sequence of eight different comparisons

applied to the comparator inputs (Fig. 5.2d). Neglecting mismatch effects, it is possible to observe that comparator changes its state with nearly zero systematic offset (and it does not suffer from any visible residual hysteresis). As a second test and, in order to validate the speed and functionality, a differential signal with a sequence of eight possible worst-case inputs, shown in Fig. 5.4 (mid.), has been applied.

As it can be observed, the proposed comparator always decides correctly, within a worst-case regeneration-time below 10 μs. A total comparison-time of 20 μs (corresponding to f_{clk} = 50 kHz) has been used, targeting an accuracy better than 10 mV. Notice that the digital outputs (Q and NQ) are valid in the falling-edge of *clk* (after half clock-cycle delay).

Table 5.2 summarizes the simulated key performance parameters of the proposed comparator, assuming a nominal V_{on} close to 0 V.

Table 5.2 Simulated key performance parameters

Characteristic	Value
Nominal supply voltage	10 V
Total pre-amp DC gain	18 dB
Simulated accuracy	10 mV$_{pp\text{-diff}}$
Comparison time	20 μs
Comparator's nominal bias current	50 μA
Static and dynamic current consumption (average value)	380 μA

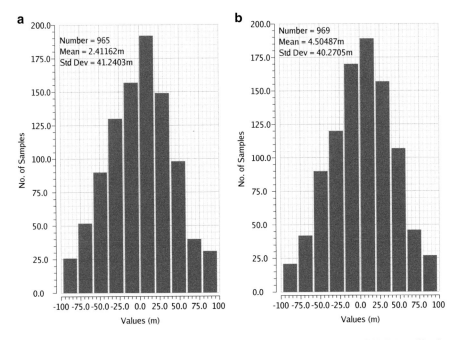

Fig. 5.5 Simulated 1-σ random offset voltage (σ(V$_{on}$ = 50 mV)) for: (**a**) V$_{on}$ = 0 V @ I$_B$ = 50 μA; (**b**) V$_{on}$ = 0.5 V @ I$_B$ = 30 μA

To double-check robustness against V$_{on}$ variations verified previously for sputtered IGZO TFTs, the proposed comparator has been also simulated using different TFT models varying V$_{on}$ between −1 and 1 V. Despite being necessary to adjust the bias current, the comparator presents always a similar behavior than when V$_{on}$ =0 V.

Random variations of the differences in the V$_{on}$ produce device mismatches, which limit the accuracy of this type of comparators. To simulate its behavior, a σ(V$_{on}$) = 50 mV has been used in 1000-cases Monte-Carlo simulations. Figure 5.5 shows the simulated 1-σ random offset voltage σ(V$_{os}$) for different V$_{on}$ values (to illustrate the robustness of the circuit) and in the positive slope of the triangular signal.

Table 5.3 Simulated comparator's offset with different V_{on}

	$V_{on} = 0$ V @ $I_B = 50$ μA		$V_{on} = 0.5$ V @ $I_B = 30$ μA	
	mV	% of full-scale	mV	% of full-scale
1-σ random offset voltage, V_{os}	<42	<0.42	<41	<0.41
1-σ random offset voltage, V_{os} (with autozeroing)	<11	<0.11	<9	<0.09

As it can be observed in Fig. 5.5 and Table 5.3, the proposed comparator achieves a $\sigma(V_{os})$ smaller than 42 mV. Additionally, employing the conventional auto-zeroing techniques this value can be reduced to less than 10 mV (by nulling the offset of the three pre-amps).

In briefly, this comparator presents low offset and the simulation results show that it is able to work at several tens of kHz, with an accuracy of the order of 10 mV. Furthermore, it supports V_{on} variations and it tolerates a reasonable level of mismatch, which are excellent properties taking into account the TFT technology involved and the variations previously studied.

5.3 ΣΔ Modulator: Circuit and Simulation Results

The modulator has been designed taking into account the fully differential second-order continuous time (CT) ΣΔM proposed for biomedical applications and experimentally evaluated at CTS/UNINOVA [3, 4] in deep nanoscale CMOS. As a standard procedure and as stated before, it will be assumed here that the digital decimation filter will be implemented in the digital-signal processor (DSP) available in the targeted application.

Figure 5.6b shows the differential second-order CT ΣΔM used in this work. The analog loop filter is composed by a cascade connection of two fully passive RC-type integrators, which are responsible to implement the second-order noise shaping transfer function (after closing the loop). In fact, it has two poles and, for a maximum attenuation of the quantization noise, they should be close to zero. Furthermore, a zero is also added by the insertion of R_5, to stabilize the loop. Some considerations about the obtaining transfer function of the circuit can be found in [5].

Despite suitable results obtained, when the previously comparator was integrated in ΣΔM, some adjustments were done:

- The channel length of digital blocks was reduced to 10 μm (considered a safe limit for production) to improve the speed;

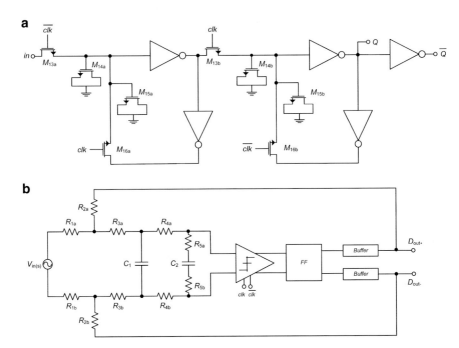

Fig. 5.6 Schematic differential second-order CT ΣΔM with feedforward structure: transmission gate D-type flip-flop (**a**) and the modulator (**b**)

- The sizing (aspect ratio) of devices $M_{1a,b}$ and multiplicity of $M_{2a,b}$ have been also increased in order to obtain a higher pre-amp DC gain. Table 5.4 shows the final sizing and multiplicity of all devices;
- The number of cascaded output inverters has been also increased for eight to properly regenerate the output and, simultaneously, increase (smoothly) the output driving capability.

The sizing of passive elements, presented in Table 5.5, was optimized through a genetic algorithm (GA) tool implemented in MATLAB/SIMULINK® in order to improve the modulator's overall dynamic performance [6]. This algorithm takes into account all characteristics such as the stability required for circuit and the comparator's delay, which affects directly the modulator. The exact transfer function was also analyzed by tool to evaluate the circuit behavior.

The final comparator design relies on the previously described topology (Fig. 5.2d) after some additional fine sizing adjustments, by adding an extra D-type flip-flop (Fig. 5.6a) and two extra output digital buffers (able to drive the output pads). Regarding the flip-flop, the inverters sizing is the same than for inverters used in the final comparator block and the remaining devices are sized with 40/10 (μm/μm) using a multiplicity of 2. The buffer presents the same architecture than inverter but using a different sizing the load and driver devices have an aspect ratio of 160/10 (μm/μm) using a multiplicity of 4 and 8, respectively. The goal

Table 5.4 Transistors' dimensions used in the final sizing of the comparator

Transistor	Characteristics		
	Width (μm)	Length (μm)	Multiplicity
$M_{1a,b}$	40	40	2
$M_{2a,b}$	160	20	12
$M_{3a,b}$	40	10	2
$M_{4a,b}$	160	10	2
$M_{5a,b}$	40	10	4
$M_{6a,b}$	40	10	8
$M_{7a,b}$	40	10	8
$M_{8a,b}$	160	10	2
M_9	40	10	2
M_{10}	160	10	2
$M_{11a,b}$	40	10	2
$M_{12a,b}$	40	10	2

Table 5.5 Sizing of passive elements obtained by MATLAB/SIMULINK®

Component	Value
$M_{1a,b}$	111.0 kΩ
$M_{2a,b}$	64.4 kΩ
$M_{3a,b}$	116.3 kΩ
$M_{4a,b}$	271.6 kΩ
$M_{5a,b}$	25.05 kΩ
C_1	2.2 nF
C_2	2.2 nF

of the flip-flop and buffering blocks is mainly to reduce the influence of jitter noise. In fact, they guarantee a constant delay which means that the logic value is always sampling at the same instant. The usage of buffers is also related to the transformation of electrical impedance between sub-circuits (and proper driving capability).

The main characteristics of ΣΔM previously defined and obtained by MATLAB/SIMULINK® tool and by Cadence are shown in Table 5.6.

A fast Fourier transform (FFT) has been obtained through an electrical simulation and it is shown in Fig. 5.7. It exhibits the slope of 40 dB/dec, compatible with a second-noise shaping order system. The complete electrical simulation has been performed using a transient-noise option, and the main results are summarized in Fig. 5.8 (for different input signal amplitudes). A simulated peak SNDR close to 57 dB has been obtained. A small difference is verified when compared to the estimated SNDR by the GA tool, probably, due to the jitter noise influence (generated internally by the digital clock buffering circuitry). A peak DR of 65 dB was also verified by electrical simulation. In terms of effective number of bits (ENOB), the peak SNDR and the peak DR are compatible with 9 and 10.5 bits, respectively.

Table 5.6 Simulated main performance parameters of the ΣΔM

Characteristic	Value
Sampling frequency (F_s)	128 kHz
Input signal amplitude (A_{in})	5 V$_{diff}$
Input signal frequency (F_{in})	101.5625 Hz
Oversampling ratio (OSR)	128
Number of points	2^{14}
Bandwidth (BW)	500 Hz
Signal-to-noise distortion ratio (SNDR)	62 (estimated) | 57 (obtained) dB
Dynamic range (DR)	65 dB
Power supply (V_{DD})	10 V
Power dissipation	22 mW

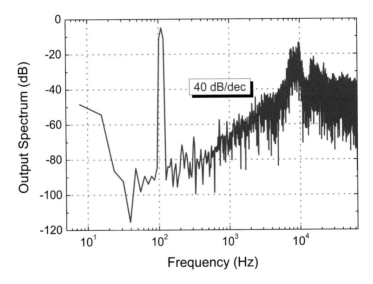

Fig. 5.7 Simulated FFT assuming a −6 dBFS and 100 Hz input signal

Regarding the usage of IGZO TFTs in complex circuits, it is clear that the operation frequency and intrinsic gain of these devices limit the performance of the ADC, being necessary the usage of high quantity of devices, thus increasing the probability of device mismatch which can be extremely crucial in other ADCs architectures. These topics are directly related to TFT devices' performance suggesting that a continuous optimization of devices is mandatory, improving not only the materials used but also the configurations and production techniques. However, when compared the results with deep nanoscale CMOS technologies, at relatively low frequencies, the performance is compatible, except mainly in terms of power dissipation and, of course, in terms of active (die) area (since TFTs fit into the category of "large area electronics") being an excellent evidence that improving devices better results can potentially be achieved.

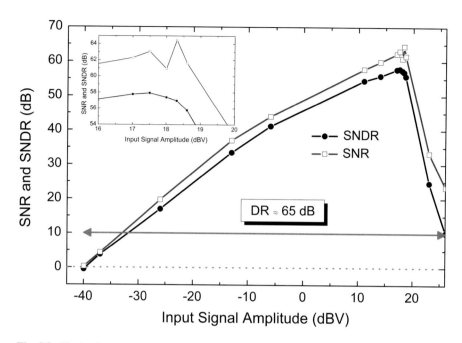

Fig. 5.8 Obtained output spectrum obtained from electrical simulation of the complete ΣΔM (SNDR versus input signal amplitude and SNR versus input signal amplitude)

Taking into account all simulation results, and when compared to the results for ADCs using different thin-film technologies (Table 5.7), it is clear that in some cases, they are above the current state-of-the-art for organics or even LTPS (either in terms of power dissipation or in reachable dynamic linearity, i.e., SNDR, DR, and ENOB). However, it is relevant to understand that the previous "comparison" is between different technologies that have associated other device performances and mechanisms. Despite the ADCs' results shown in Table 5.7 were obtained after fabrication, the simulation results achieved in this work are encouraging and give an excellent perspective to produce the first second-order ΣΔ ADCs using oxide TFTs.

5.4 Circuit Layout

After schematic-level simulations, the circuit layout is considered a paramount step in ICs. Basically, the main intention is to draw the lithographic masks which will be used in fabrication process. Given that for these specific devices produced in clean room there is no process design kit (PDK) available it was suggested the creation of a PCELL in order to simplify the future layout. This PCELL was constructed taking into account the staggered bottom gate, top contact structure, previously

Table 5.7 Comparison between this work and the most relevant published ADCs using thin-film technologies

Technology	Architecture	Year	TFT characteristics	ADC characteristics	Authors
a-Si:H (n-type)	Flash (2 kS/s)	2012	Bottom-gate inverted staggered structure Low temperature:180 °C $V_T = 1.2$ V $\mu_{sat} = 0.8$ cm²/V s On/Off $= 10^7$	5 bits Without calibration $V_{DD} = 10$ V Power consumption: 13.6 mW	Dey and Allee [7]
Poly-Si TFTs (laser crystallized)	Flash (3 MS/s)	2010	$\mu_{n\text{-type}} = 281$ cm²/V s $\mu_{p\text{-type}} = 98$ cm²/V s Process: PECVD + sequential lateral solidification (SLS)	3 bits Stainless steel foil substrate	Jamshidi-Roudbari et al. [8]
LTPS TFTs	Second-order $\Sigma\Delta$	2009	–	$V_{DD} = 11.2$ V DR $= 69$ dB SNDR $= 65.63$ dB Power consumption: 63.3 mW	Lin et al. [9]
Complementary organic TFTs	SAR ($F_S = 100$ Hz)	2010	$\mu_{n\text{-type}} = 0.02$ cm²/V s $\mu_{p\text{-type}} = 0.5$ cm²/V s	6 Bits Power consumption: 3.6 μW	Xiong et al. [10]
Organic TFTs	First-order $\Sigma\Delta$ (OSR=16)	2011	Pentacene-based dual-gate organic TFTs	Flexible plastic foil SNR $= 26.5$ dB $V_{DD} = 15$ V Power consumption: 1.5 mW	Marien et al. [2]
Oxide TFTs (IGZO TFTs)	Second-order $\Sigma\Delta$	2014/2015	Staggered bottom-gate, top-contact structure $V_{on} = -1$ V $\mu_{FE} > 14$ cm²/V s On/Off $= 10^8$ S $= 0.16$ V/decade	$V_{DD} = 10$ V DR $= 65$ dB SNDR $= 57$ dB Power consumption: 22 mW Die area: 10 mm²	This work[a]

[a] The presented performance parameters for the ADC proposed in this work were obtained through electrical simulation

Table 5.8 Rules used during circuit layout

Description	
Minimum space between metals	$20\,\mu m$
Minimum width of metals	$20\,\mu m$ (except in channel)
Minimum area of vias	$20 \times 20\,\mu m^2$
Minimum space between vias	$20\,\mu m$
Overlap gate—contacts	$2\,\mu m$
Overlap semiconductor—contacts	$4\,\mu m$
Gate extension	$5\,\mu m$
Contacts extension	$2.5\,\mu m$

used to fabricate individual devices. The layers and overlapping dimensions are presented in Table 5.8. These rules take into account the limitation of the equipment and processes established in CENIMAT, being a worst-case scenario considering relevant misalignment and a safety factor to prevent undesirable effects.

Since the circuit presents passive and active elements, and considering that the project is at an early stage, it was decided that the passive elements would be realized using surface mounted devices (SMDs). These devices will be either glued in the glass substrate or placed in a testing printed circuit board (PCB), in order to have more degrees of freedom to optimize during the production process. After that, the main idea is to produce the complete circuit in a glass substrate, modifying the mask set for resistors and capacitors fabrication. However, the sputtering process for these components needs to be optimized essentially in terms of mismatch. The connection between glass and PCB will be done using a commercial connector (connector PCB card edge), where an edge of the glass has to be abraded to insert it in the connector (Fig. 5.9). This method to link the entire circuit is a simple way to avoid wire bonding process and, at the same time, the direct contact with sample, preventing some possible electrostatic discharge (ESD).

Considering just the active elements of the circuit, its complexity is particularly high. In this sense, it was extremely relevant to think in different aspects that can restrict (limit) the operation of the circuit. For instance, parasitic capacitances effects were prevented due to the inclusion of a dielectric layer of parylene between metal layers. This dielectric was chosen by virtue of its low-κ and of the possibility to produce a thick film ($\approx 1\,\mu m$). The minimum overlap and the maximum length of lines were also considered in order to avoid possible antenna effects. As a consequence, a set composed by 7 masks was designed:

- Metal 1 deposition (gates) and lift-off (**mask 1**);
- Dielectric deposition (multilayer structure);
- Semiconductor deposition and lift-off (**mask 2**);
- Parylene deposition and etching to open vias (**mask 3**);
- Etching of multilayer dielectric to access to metal 1 in opened vias (**mask 4**);
- Metal 2 deposition and lift-off (**mask 5**);
- Deposition of contact pads and lift-off (**mask 6**);
- Passivation and etching (**mask 7**).

Fig. 5.9 Glass substrate, covered with a molybdenum thin film, inserted in a PCB connector for testing

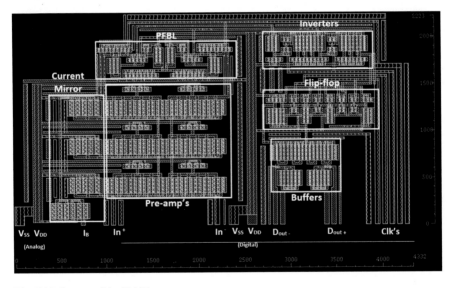

Fig. 5.10 Layout of the $\Sigma\Delta$M

Some individual transistors were also included around the substrate as test structures, allow to understanding how materials are varying depending on position in substrate, which will be relevant to comprehend the experimental results of the complete circuit.

Figure 5.10 shows the layout of circuit which has a die area close to $10\,\text{mm}^2$, including dummy structures (for both devices and metal lines). Given the absence of PDK, automatic design rule check (DRC) and layout versus schematic (LVS) cannot be performed using EDA tools. As a consequence, the layout of the entire circuit has still to be checked manually.

References

1. T.C. Carusone, D. Johns, K. Martin, *Analog Integrated Circuit Design*, 2nd ed. (Wiley, 2011). http://eu.wiley.com/WileyCDA/WileyTitle/productCd-EHEP002039.html
2. H. Marien, M.S.J. Steyaert, E. van Veenendaal, P. Heremans, A fully integrated $\Delta\Sigma$ ADC in organic thin-film transistor technology on flexible plastic foil. IEEE J. Solid State Circuits **46**(1), 276–284 (2011)
3. J.L.A. de Melo, F. Querido, N. Paulino, J. Goes, A 0.4-V 410-nW opamp-less continuous-time $\Sigma\Delta$ modulator for biomedical applications, in *2014 IEEE International Symposium on Circuits and Systems (ISCAS)* (IEEE, Melbourne, 2014) pp. 1340–1343
4. J.L.A. de Melo, J. Goes, N. Paulino, A 0.7 V 256 μW $\Delta\Sigma$ modulator with passive RC integrators achieving 76 dB DR in 2 MHz BW, in *2015 Symposium on VLSI Circuits (VLSI Circuits)* (IEEE, Kyoto, 2015) pp. C290–C291
5. J.L.A. De Melo, A low power 1-MHz continuous-time $\Sigma\Delta$M using a passive loop filter designed with a genetic algorithm tool, in *Proceedings - IEEE International Symposium on Circuits and Systems*, vol. 1 (2013), pp. 586–589
6. J.L.A. de Melo, B. Nowacki, N. Paulino, J. Goes, Design methodology for Sigma-Delta modulators based on a genetic algorithm using hybrid cost functions, in *2012 IEEE International Symposium on Circuits and Systems* (IEEE, Seoul, 2012) pp. 301–304
7. A. Dey, D.R. Allee, Amorphous silicon 5 bit flash analog to digital converter, in *2012 IEEE Custom Integrated Circuits Conference* (IEEE, San Jose, 2012)
8. A. Jamshidi-Roudbari, P.-C. Kuo, M.K. Hatalis, A flash analog to digital converter on stainless steel foil substrate. Solid-State Electron. **54**(4), 410–416 (2010)
9. W.-M. Lin, C.-F. Lin, S.-I. Liu, A CBSC second-order sigma-delta modulator in 3 μm LTPS-TFT technology, in *2009 IEEE Asian Solid-State Circuits Conference* (IEEE, Taipei, 2009) pp. 133–136
10. W. Xiong, U. Zschieschang, H. Klauk, B. Murmann, A 3V 6b successive-approximation ADC using complementary organic thin-film transistors on glass, in *2010 IEEE International Solid-State Circuits Conference - (ISSCC)* (IEEE, San Francisco, 2010) pp. 134–135

Chapter 6
Conclusions and Future Perspectives

This work involved three main topics: study and optimization of multicomponent and multilayers dielectrics and their integration in IGZO TFTs; $\Sigma\Delta$ modulator design and simulation using a model specially adapted for IGZO TFTs; circuit layout for future fabrication using optimized fabrication processes. In this section, the most relevant conclusions and future perspectives about these topics are addressed.

Regarding the study and optimization of the dielectric layer, it was verified that multicomponent and multilayer structure provides better dielectric properties, having a direct impact in IGZO TFT performance:

- In terms of single layer dielectrics, co-sputtered TSiO presented a high-κ and reasonable E_G, despite its E_B being low for fabrication of reliable devices. The TFTs produced using TSiO exhibited a considerable improvement when compared to the devices employing a Ta_2O_5 layer, mainly in terms of S and I_G. However, stress measurements revealed a quite significant negative ΔV_{on} after a 3 h stress (≈-20 V, compared to ≈-40 V of Ta_2O_5), suggesting a large ionic movement inside the dielectric layer. Nonetheless, this co-sputtered layer revealed a smooth surface and an amorphous structure, which are crucial properties for integration into multilayer configuration;
- Concerning the multilayer structures, both configurations (three and seven layers) presented good properties, with the structure using seven layers providing a very large E_B (>7.5 MV/cm), while maintaining a reasonable κ (>10). When integrated in TFTs, devices exhibited very low operating voltage ($V_{on} =-1.5$ V), low I_G (<10 pA) and the magnitude of stress mechanisms was considerably reduced when compared to TFTs with single layer dielectrics ($\Delta V_{on} \approx 1$ V). The performance of TFTs using this layer was considered appropriate for future usage in complex circuits: after annealing

© The Author(s) 2016
A.P.P. Correia et al., *A Second-Order ΣΔ ADC Using Sputtered IGZO TFTs*,
SpringerBriefs in Electrical and Computer Engineering,
DOI 10.1007/978-3-319-27192-7_6

at 200 °C, $\mu_{sat} \approx 13 \, cm^2/V \, s$, On/Off $\approx 10^7$ and S $\approx 0.2 \, V/dec$, which are similar results to IGZO TFTs employing dielectrics processed at temperatures exceeding 300 °C.[1]

Circuit design and simulation required the availability of a TFT model that describes properly the device's behavior. In this sense, the a-Si TFT model developed by Semiconductor Devices Research Group at RPI was adjusted. The I–V characteristics of IGZO TFTs are properly fitted using this model with good accuracy. The model only focused static characteristics but transients should also be included in future and upgraded versions. Still, the degree of accuracy of the present model was perfectly suitable to enable circuit design/simulation.

The design and electrical simulations of the complete $\Sigma\Delta M$ revealed quite attractive and encouraging results:

- The comparator circuit works at several tens of kHz with an accuracy of 10 mV. This circuit presented a current consumption as small as 380 μA drawn from a 10 V positive power supply. Furthermore, in order to guarantee the robustness of this comparator against V_{on} variations due to the mismatch, it was also simulated using different V_{on} between −1 and 1 V;
- Due to IGZO TFTs performance, the $\Sigma\Delta M$ comprises a relatively large quantity of devices. However, simulation results show a peak SNDR of 57 dB and a DR of 65 dB (BW =500 Hz) with a power dissipation of about 22 mW. These are considered very interesting results, going beyond the state-of-the-art when comparing with other ADC circuit implementations based on other transistor's technologies, either using organic devices or even LTPS.

In terms of circuit layout, due to the fact that there was not a PDK available and it is the first time that such a complex circuit is produced in CENIMAT, a large number of details had to be considered and studied to create a set of conservative design rules, to achieve a good yield of fabricated devices. A mask set composed by seven masks was designed, considering a die area close to 10 mm². The final layout is currently under manual verification.

Despite all work involved and the future production in clean room environment of the $\Sigma\Delta M$ ADC which is the main future perspective—some questions and ideas remain unanswered, and they will required further investigation:

- Regarding TFTs, a continuous optimization is crucial to increase the complexity and performance of circuits where they are integrated. In this sense, the influence of Ta_2O_5/SiO_2 ratio during dielectric deposition should be carefully analyzed. Furthermore, the performance variation of TFTs depending on their

---

[1]It should also be mentioned that during the last months, after the conclusion of the work described in this book, further dielectric thickness optimization allowed to reduce annealing temperature down to 180 °C without any degradation of device performance/stability.

placement in substrate should be studied in order to understand whether there is a gradient associated or even if deposition conditions can be improved. The significant variations verified during stress measurements need a particular attention and investigation in order to understand, exactly, the mechanisms involved and how they can be suppressed to not limit TFT performance;

- In terms of modeling, new approaches are now being studied by the group, expanding the current static model to also consider transients, improving the accuracy of the simulation results. This is being done both by physical and empirical models.
- After a successful fabrication of this $\Sigma\Delta M$ whose layout only includes transistors (resistors and capacitors will be implemented in a PCB), a complete $\Sigma\Delta M$ will be investigated including the passive devices in the same substrate. As a consequence, the fabrication of resistors and capacitors with acceptable passive device's mismatch will be necessary in order to guarantee a satisfactory circuit performance
- Taking into account all process, a $\Sigma\Delta M$ using a flexible and/or paper substrate, it is also a goal for future to reinforce the concept of flexible and recyclable electronics. The feasibility of other ADC architectures (e.g., Nyquist-rate ADCs) should also be further investigated.

Printed in the United States
By Bookmasters